Aerodynamics for Engineers

Aerodynamics
for Engineers

Aerodynamics for Engineers

Noah Stokes

WILLFORD PRESS

www.willfordpress.com

Published by Willford Press,
118-35 Queens Blvd., Suite 400,
Forest Hills, NY 11375, USA

ISBN: 978-1-64728-323-0

Cataloging-in-Publication Data

Aerodynamics for engineers / Noah Stokes.
p. cm.
Includes bibliographical references and index.
ISBN 978-1-64728-323-0
1. Aerodynamics. 2. Aerospace engineering. 3. Aeronautics. I. Stokes, Noah.
TL570 .A37 2022
620.107.4--dc23

For information on all Willford Press publications
visit our website at www.willfordpress.com

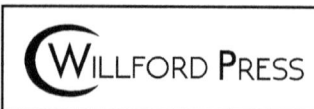

WILLFORD PRESS

Contents

Permissions

Index

Preface

The field of physics, which studies the motion of air, primarily in conjunction with its interactions with a solid object is known as aerodynamics. Some of the forces, which are studied within this discipline are the forces of flight: drag and lift. However, the forces of thrust and weight may also be studied in aerodynamic problems. Air is generally assumed to behave as a continuum, rather than as discrete particles. This allows the problems in aerodynamics to be solved using fluid dynamic conservation laws. The principles behind these laws are conservation of mass, conservation of energy and conservation of momentum. This book provides significant information of this discipline to help develop a good understanding of aerodynamics and related fields. It will serve as a valuable source of reference for graduate and post graduate students. For all those who are interested in this field, this book can prove to be an essential guide.

A detailed account of the significant topics covered in this book is provided below:

Chapter 1- Aerodynamics is defined as the study of motion of air and its interaction with a solid object. The forces that fall under its domain are lift, drag, thrust and weight. The topics elaborated in this chapter will help in gaining a better perspective about these forces of aerodynamics.

Chapter 2- Some of the fundamental principles and equations of aerodynamics include conservation of energy, compressible and incompressible flow, aerodynamic force and circulation, velocity equation, momentum equation, etc. All these principles and equations of aerodynamics have been carefully analyzed in this chapter.

Chapter 3- The branch of physics that deals with the study of interaction between the inertial, elastic, and aerodynamic forces is defined as aeroelasticity. A few of its types are static aeroelasticity, dynamic aeroelasticity, nonlinear aeroelasticity, etc. This chapter discusses these types of aeroelasticity in detail.

Chapter 4- A propeller is a device that transforms rotational power into linear thrust by creating a pressure between the two surfaces and acting upon a working fluid such as water or air. Airfoil aerodynamics refers to the cross-sectional shape of a wing that is used in various aeronautical applications. The diverse aspects of propeller and airfoil aerodynamics have been thoroughly discussed in this chapter.

Chapter 5- The flight instruments depend on direct measurement of aerodynamic pressure to predict the altitude, airspeed and climb rate of the aircraft. It includes barometric altimeter, radio altimeter, vertical speed indicator, pilot-static probe, etc. This chapter closely examines these measurement tools of aerodynamic pressure to provide an extensive understanding of the subject.

It gives me an immense pleasure to thank our entire team for their efforts. Finally in the end, I would like to thank my family and colleagues who have been a great source of inspiration and support.

Noah Stokes

Aerodynamics: An Introduction

Aerodynamics is defined as the study of motion of air and its interaction with a solid object. The forces that fall under its domain are lift, drag, thrust and weight. The topics elaborated in this chapter will help in gaining a better perspective about these forces of aerodynamics.

Aerodynamics is a branch of fluid dynamics concerned with studying the principles of the flow of gases and the forces generated on a solid body within the flow. To solve a problem in aerodynamics, the researcher needs to take into account various properties of the flow, such as velocity, pressure, density, and temperature, as functions of space and time. By understanding the flow pattern, it becomes possible to calculate or estimate the forces and moments acting on solid bodies within the flow.

Aerodynamics forms the scientific basis for heavier-than-air flight. In addition, it plays an important role in the design of automobiles, large buildings, bridges, ventilation passages, and hard drive heads. Town planners need to consider aerodynamics to reduce pollution and increase comfort in outdoor spaces, and environmental scientists study aerodynamics to understand air circulation and how flight mechanics affect ecosystems.

A vortex is created by the passage of an aircraft wing, revealed by colored smoke. Vortices are one of the many phenomena associated to the study of aerodynamics. The equations of aerodynamics show that the vortex is created by the difference in pressure between the upper and lower surface of the wing. At the end of the wing, the higher pressure on the lower surface effectively tries to 'reach over' to the low pressure side, creating rotation and the vortex.

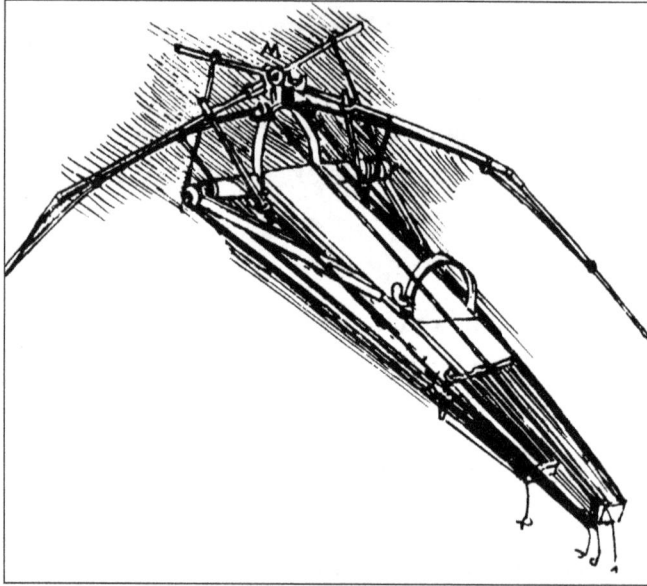

A drawing of a design for a flying machine by Leonardo da Vinci.
This machine was an ornithopter, similar to a modern day helicopter.

Images and stories of flight have appeared throughout recorded history. One of the most notable of these is the story of Icarus and Daedalus. Although observations of some aerodynamic effects, such as wind resistance (or drag) were recorded by the likes of Aristotle and Galileo Galilei, very little effort was made to develop governing laws for understanding the nature of flight prior to the seventeenth century.

Sir Isaac Newton was the first person to develop a theory of air resistance in 1726, arguably making him the world's first aerodynamicist. As part of that theory, Newton believed that drag was due to the dimensions of a body, the density of the fluid, and the velocity raised to the second power. These ideas all turned out to be correct for low-flow speeds. Newton also developed a law for the drag force on a flat plate inclined toward the direction of the fluid flow. Using F for the drag force, ρ for the density, S for the area of the flat plate, V for the flow velocity, and θ for the inclination angle, his law was expressed by the following equation:

$$F = \rho S V^2 \sin^2\left(\theta\right)$$

Unfortunately, this equation is entirely incorrect for the calculation of drag (unless the flow speed is hypersonic). Drag on a flat plate is closer to being linear with the angle of inclination as opposed to acting quadratically. This formula can lead one to believe that flight is more difficult than it actually is, and it may have contributed to a delay in manned flight.

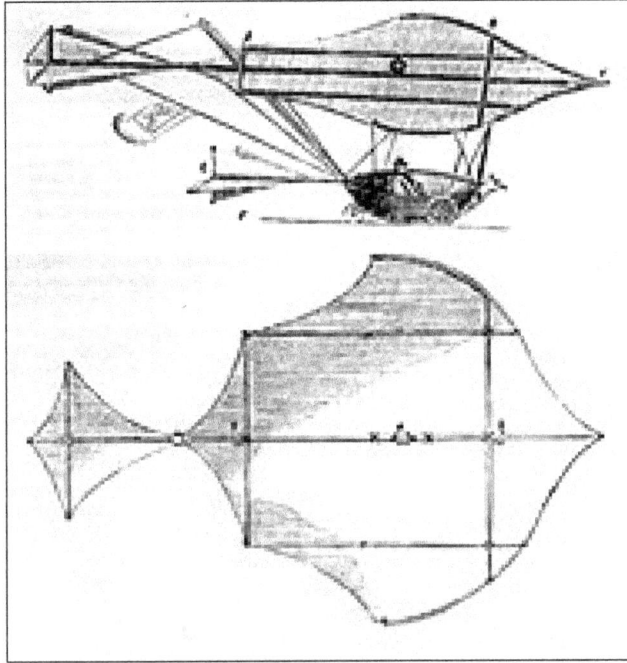

A drawing of a glider by Sir George Cayley, one of the
early attempts at creating an aerodynamic shape.

Sir George Cayley is credited as the first person to separate the forces of lift and drag
which are in effect on any flight vehicle. Cayley believed that the drag on a flying machine must be counteracted by a means of propulsion in order for level flight to occur.
Cayley also looked to nature for aerodynamic shapes with low drag. One of the shapes
he investigated were the cross-sections of trout. The bodies of fish are shaped to produce very low resistance as they travel through water. As such, their cross-sections are
sometimes very close to that of modern low-drag airfoils.

These empirical findings led to a variety of air resistance experiments on various shapes
throughout the eighteenth and nineteenth centuries. Drag theories were developed by
Jean le Rond d'Alembert, Gustav Kirchhoff, and Lord Rayleigh. Equations for fluid flow
with friction were developed by Claude-Louis Navier and George Gabriel Stokes. To
simulate fluid flow, many experiments involved immersing objects in streams of water
or simply dropping them off the top of a tall building. Towards the end of this time
period, Gustave Eiffel used his Eiffel Tower to assist in the drop testing of flat plates.

A more precise way to measure resistance is to place an object within an artificial, uniform stream of air where the velocity is known. The first person to experiment in this
fashion was Francis Herbert Wenham, who in doing so constructed the first wind tunnel in 1871. Wenham was also a member of the first professional organization dedicated
to aeronautics, the Royal Aeronautical Society of Great Britain. Objects placed in wind
tunnel models are almost always smaller than in practice, so a method was needed to
relate small scale models to their real-life counterparts. This was achieved with the

invention of the dimensionless Reynolds number by Osborne Reynolds. Reynolds also experimented with laminar to turbulent flow transition in 1883.

By the late nineteenth century, two problems were identified before heavier-than-air flight could be realized. The first was the creation of low-drag, high-lift aerodynamic wings. The second problem was how to determine the power needed for sustained flight. During this time, the groundwork was laid down for modern day fluid dynamics and aerodynamics, with other less scientifically inclined enthusiasts testing various flying machines with little success.

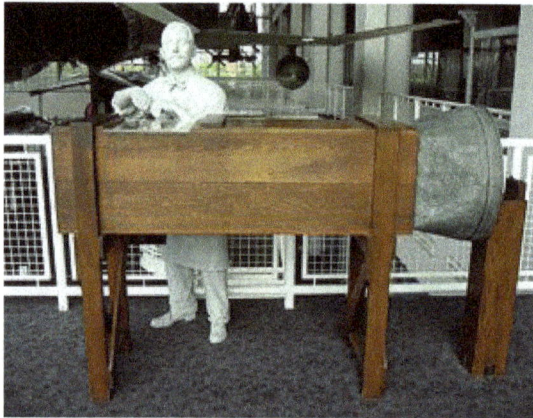

A replica of the Wright Brothers' wind tunnel is on display at the Virginia Air and Space Center. Wind tunnels were key in the development and validation of the laws of aerodynamics.

In 1889, Charles Renard, a French aeronautical engineer, became the first person to reasonably predict the power needed for sustained flight. Renard and German physicist Hermann von Helmholtz explored the wing loading of birds, eventually concluding that humans could not fly under their own power by attaching wings onto their arms. Otto Lilienthal, following the work of Sir George Cayley, was the first person to become highly successful with glider flights. Lilienthal believed that thin, curved airfoils would produce high lift and low drag.

Octave Chanute provided a great service to those interested in aerodynamics and flying machines by publishing a book outlining all of the research conducted around the world up to 1893. With the information contained in that book and the personal assistance of Chanute himself, the Wright brothers had just enough knowledge of aerodynamics to fly the first manned aircraft on December 17, 1903, just in time to beat the efforts of Samuel Pierpont Langley. The Wright brothers' flight confirmed or disproved a number of aerodynamics theories. Newton's drag force theory was finally proved incorrect. The first flight led to a more organized effort between aviators and scientists, leading the way to modern aerodynamics.

During the time of the first flights, Frederick W. Lanchester, Martin Wilhelm Kutta, and Nikolai Zhukovsky independently created theories that connected circulation of a fluid flow to lift. Kutta and Zhukovsky went on to develop a two-dimensional wing theory. Expanding upon the work of Lanchester, Ludwig Prandtl is credited with developing the mathematics behind thin-airfoil and lifting-line theories as well as work with boundary layers. Prandtl, a professor at Gottingen University, instructed many students who would play important roles in the development of aerodynamics like Theodore von Kármán and Max Munk.

As aircraft began to travel faster, aerodynamicists realized that the density of air began to change as it came into contact with an object, leading to a division of fluid flow into the incompressible and compressible regimes. In compressible aerodynamics, density and pressure both change, which is the basis for calculating the speed of sound. Newton was the first to develop a mathematical model for calculating the speed of sound, but it was not correct until Pierre-Simon Laplace accounted for the molecular behavior of gases and introduced the heat capacity ratio. The ratio of the flow speed to the speed of sound was named the Mach number after Ernst Mach, who was one of the first to investigate the properties of supersonic flow which included Schlieren photography techniques to visualize the changes in density. William John Macquorn Rankine and Pierre Henri Hugoniot independently developed the theory for flow properties before and after a shock wave. Jakob Ackeret led the initial work on calculating the lift and drag on a supersonic airfoil. Theodore von Kármán and Hugh Latimer Dryden introduced the term transonic to describe flow speeds around Mach 1 where drag increases rapidly. Because of the increase in drag approaching Mach 1, aerodynamicists and aviators disagreed on whether manned supersonic flight was achievable.

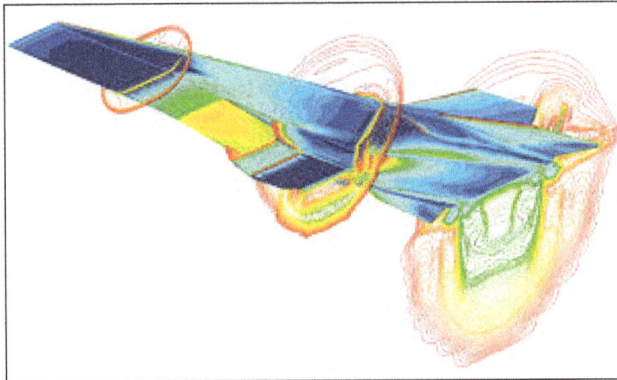

A computer generated model of NASA's X-43A hypersonic research vehicle flying at Mach 7 using a computational fluid dynamics code.

On September 30, 1935 an exclusive conference was held in Rome with the topic of high velocity flight and the possibility of breaking the sound barrier. Participants included von Kármán, Prandtl, Ackeret, Eastman Jacobs, Adolf Busemann, Geoffrey Ingram Taylor, Gaetano Arturo Crocco, and Enrico Pistolesi. The new research presented was impressive. Ackeret presented a design for a supersonic wind tunnel. Busemann gave perhaps the best presentation on the need for aircraft with swept wings for high speed flight. Eastman Jacobs, working for NACA, presented his optimized airfoils for high subsonic speeds which led to some of the high performance American aircraft during World War II.

By the time the sound barrier was broken, much of the subsonic and low supersonic aerodynamics knowledge had matured. The Cold War fueled an ever evolving line of high performance aircraft. Computational fluid dynamics was started as an effort to solve for flow properties around complex objects and has rapidly grown to the point where entire aircraft can be designed using a computer.

With some exceptions, the knowledge of hypersonic aerodynamics has matured between the 1960s and the present decade. Therefore, the goals of an aerodynamicist have shifted from understanding the behavior of fluid flow to understanding how to engineer a vehicle to interact appropriately with the fluid flow. For example, while the behavior of hypersonic flow is understood, building a scramjet aircraft to fly at hypersonic speeds has met with very limited success. Along with building a successful scramjet aircraft, the desire to improve the aerodynamic efficiency of current aircraft and propulsion systems will continue to fuel new research in aerodynamics.

Aerodynamic Problems: Classification

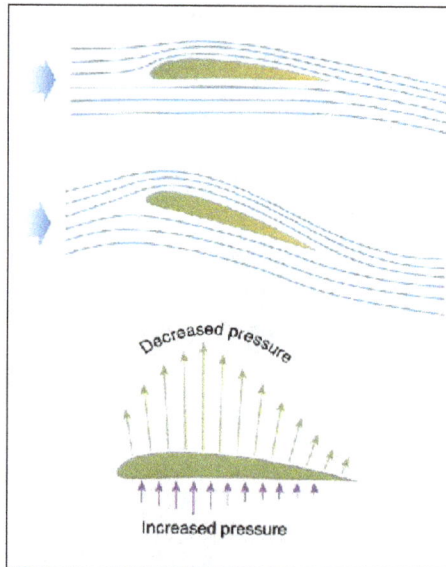

The flow of air past the wing of an aircraft.

Aerodynamic problems can be classified in a number of ways. The flow environment defines the first classification criterion. *External* aerodynamics is the study of flow around solid objects of various shapes. Examples of external aerodynamics include evaluation of the lift and drag on an airplane, the shock waves that form in front of a rocket's nose, or the flow of air over a hard drive head. *Internal* aerodynamics is the study of flow through passages in solid objects. Internal aerodynamics encompasses, for instance, the study of airflow through a jet engine or through an air conditioning pipe.

The ratio of the fluid's characteristic flow speed to the speed of sound constitutes a second type of classification of aerodynamic problems. A problem is called subsonic if all the speeds in the problem are less than the speed of sound; transonic if flow speeds both below and above the speed of sound are present (normally when the characteristic speed is approximately the speed of sound); supersonic when the characteristic flow speed is greater than the speed of sound; and hypersonic when the flow speed is much greater than the speed of sound. Aerodynamicists disagree over the precise definition of hypersonic flow; minimum Mach numbers for hypersonic flow range from 3 to 12. Most aerodynamicists use Mach numbers between 5 and 8.

The influence of viscosity in the flow dictates a third type of classification. Some problems involve negligible viscous effects on the solution, in which case viscosity may be considered nonexistent. Approximations to these problems are called inviscid flows. Flows for which viscosity cannot be neglected are called viscous flows.

Continuity Assumption

Gases are composed of molecules that collide with one another and solid objects. If density and velocity are taken to be well-defined at infinitely small points, and they are assumed to vary continuously from one point to another, the discrete molecular nature of a gas is ignored.

The continuity assumption becomes less valid as a gas becomes increasingly rarefied. In these cases, statistical mechanics is a more valid method of solving the problem than aerodynamics.

Conservation Laws

Aerodynamic problems are solved using the conservation laws, or equations derived from the conservation laws. In aerodynamics, three conservation laws are used:

- Conservation of mass: Matter is not created or destroyed. If a certain mass of fluid enters a volume, it must either exit the volume or increase the mass inside the volume.

- Conservation of momentum: This is also called Newton's second law of motion.

- Conservation of energy: Although energy can be converted from one form to another, the total energy in a given system remains constant.

Incompressible Aerodynamics

An incompressible flow is characterized by a constant density despite flowing over surfaces or inside ducts. A flow can be considered incompressible as long as its speed is low. For higher speeds, the flow will begin to compress as it comes into contact with surfaces. The Mach number is used to distinguish between incompressible and compressible flows.

Subsonic Flow

Subsonic (or low-speed) aerodynamics is the study of inviscid, incompressible and irrotational aerodynamics where the differential equations used are a simplified version of the governing equations of fluid dynamics. It is a special case of Subsonic aerodynamics.

In solving a subsonic problem, one decision to be made by the aerodynamicist is whether to incorporate the effects of compressibility. Compressibility is a description of the amount of change of density in the problem. When the effects of compressibility on the solution are small, the aerodynamicist may choose to assume that density is constant. The problem is then an incompressible low-speed aerodynamics problem. When the density is allowed to vary, the problem is called a compressible problem. In air, compressibility effects are usually ignored when the Mach number in the flow does not exceed 0.3 (about 335 feet per second or 228 miles per hour or 102 meters per second at 60 °F). Above 0.3, the problem should be solved using compressible aerodynamics.

Compressible Aerodynamics

According to the theory of aerodynamics, a flow is considered to be compressible if its change in density with respect to pressure is non-zero along a streamline. In short, this means that, unlike incompressible flow, changes in density must be considered. In general, this is the case where the Mach number in part or all of the flow exceeds 0.3. The Mach .3 value is rather arbitrary, but it is used because gas flows with a Mach number below that value demonstrate changes in density with respect to the change in pressure of less than 5%. Furthermore, that maximum 5% density change occurs at the stagnation point of an object immersed in the gas flow and the density changes around the rest of the object will be significantly lower. Transonic, supersonic, and hypersonic flows are all compressible.

Transonic Flow

The term Transonic refers to a range of velocities just below and above the local speed of sound (generally taken as Mach 0.8–1.2). It is defined as the range of speeds between the critical Mach number, when some parts of the airflow over an aircraft become supersonic, and a higher speed, typically near Mach 1.2, when all of the airflow is supersonic. Between these speeds some of the airflow is supersonic, and some is not.

Supersonic Flow

Supersonic aerodynamic problems are those involving flow speeds greater than the speed of sound. Calculating the lift on the Concorde during cruise can be an example of a supersonic aerodynamic problem.

Supersonic flow behaves very differently from subsonic flow. Fluids react to differences in pressure; pressure changes are how a fluid is "told" to respond to its environment. Therefore, since sound is in fact an infinitesimal pressure difference propagating through a fluid, the speed of sound in that fluid can be considered the fastest speed that "information" can travel in the flow. This difference most obviously manifests itself in the case of a fluid striking an object. In front of that object, the fluid builds up a stagnation pressure as impact with the object brings the moving fluid to rest. In fluid traveling at subsonic speed, this pressure disturbance can propagate upstream, changing the flow pattern ahead of the object and giving the impression that the fluid "knows" the object is there and is avoiding it. However, in a supersonic flow, the pressure disturbance cannot propagate upstream. Thus, when the fluid finally does strike the object, it is forced to change its properties — temperature, density, pressure, and Mach number — in an extremely violent and irreversible fashion called a shock wave. The presence of shock waves, along with the compressibility effects of high-velocity fluids, is the central difference between supersonic and subsonic aerodynamics problems.

Hypersonic Flow

In aerodynamics, hypersonic speeds are speeds that are highly supersonic. In the 1970s, the term generally came to refer to speeds of Mach 5 (5 times the speed of sound) and above. The hypersonic regime is a subset of the supersonic regime. Hypersonic flow is characterized by high temperature flow behind a shock wave, viscous interaction, and chemical dissociation of gas.

Associated Terminology

The incompressible and compressible flow regimes produce many associated phenomena, such as boundary layers and turbulence.

Boundary Layers

The concept of a boundary layer is important in many aerodynamic problems. The viscosity and fluid friction in the air is approximated as being significant only in this thin layer. This principle makes aerodynamics much more tractable mathematically.

Turbulence

In aerodynamics, turbulence is characterized by chaotic, stochastic property changes in the flow. This includes low momentum diffusion, high momentum convection, and rapid variation of pressure and velocity in space and time. Flow that is not turbulent is called laminar flow.

Applications of Aerodynamics in Diverse Fields

Aerodynamics is important in a number of applications other than aerospace engineering. For instance, it is a significant factor in any type of vehicle design, including automobiles. It is important in the prediction of forces and moments in sailing. It is used in the design of small components such as hard drive heads.

Structural engineers use aerodynamics, particularly aeroelasticity, to calculate wind loads in the design of large buildings and bridges. Urban aerodynamics helps town planners and designers improve comfort in outdoor spaces, create urban micro climates, and reduce the effects of urban pollution. The field of environmental aerodynamics studies the ways atmospheric circulation and flight mechanics affect ecosystems. The aerodynamics of internal passages is important in heating/ventilation, gas piping, and in automotive engines, where detailed flow patterns strongly affect the performance of the engine.

Angle of Attack

Angle of Attack is used to define the angle between the wing chord line and the flight path. This is not to be confused with the relation of the aircraft to the Earth's surface. This is called the attitude and is seldom, if ever, the same as the angle of attack.

When the angle is small, the aircraft is said to be at a low angle of attack. When the angle is large, the aircraft is said to be at a high angle of attack. During landing, an airplane may have a level attitude, but a high angle of attack, because the flight path is downward and the approaching wind is parallel to the flight path. During climb, an airplane can be in a nose-high attitude, but at a low angle of attack.

Two variables can change the amount of lift generated by a wing in a given configuration:

- The speed of air flowing over the airfoil.

- The angle of attack.

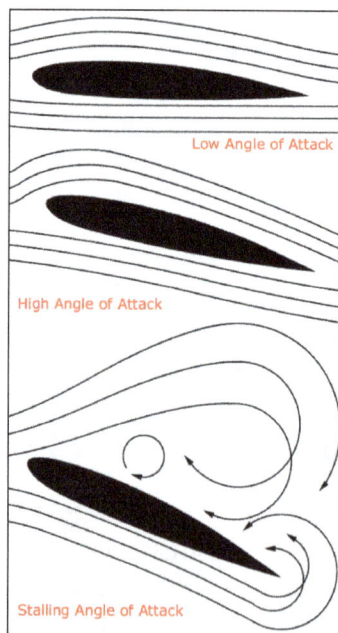

Low Angle of Attack

High Angle of Attack

Stalling Angle of Attack

An increase in speed or the angle of attack will increase both lift and drag. In level flight, lift must equal the weight of the aircraft. If an aircraft weighs 2,000 lbs., the wing must generate 2,000 lbs. of lift. The speed and angle of attack are interchangeable to a point—therefore, for every airspeed, there is a corresponding angle of attack that will produce the same amount of lift.

In order for a wing to produce lift, the air flowing past an aircraft, must be aligned to the airfoil in order to provide a smooth airflow. As a wing increases its angle of attack, airflow can no longer flow smoothly over the wing and eddies or burbles will form, causing the wing to approach its stall speed. When a wing finally stalls, it will no longer produce lift and with weight unopposed by lift, the aircraft will drop towards the ground. With sufficient altitude, stall recovery can be obtained by decreasing the angle of attack.

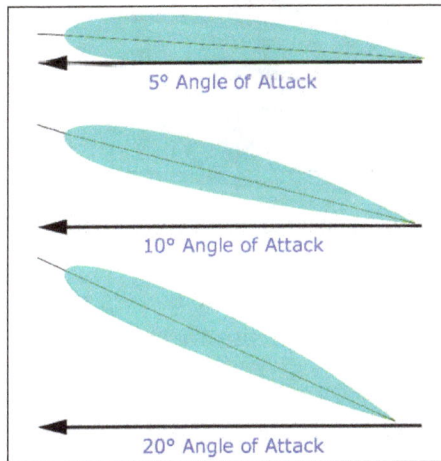

5° Angle of Attack

10° Angle of Attack

20° Angle of Attack

Angle of Attack is the angle between the wing chord line and the flight path.

An airplane can be stalled at any attitude and at any airspeed such as pulling out of a dive too abruptly or if the airplane is in a steep turn at a high angle of attack, even though the airspeed is high. During dives or turns, centrifugal force will increase load factors and if excessive, this will cause the wing to stall.

Drag

Drag is the aerodynamic force that opposes an aircraft's motion through the air. Drag is generated by every part of the airplane (even the engines).

Drag is a mechanical force. It is generated by the interaction and contact of a solid body with a fluid (liquid or gas). It is not generated by a force field, in the sense of a gravitational field or an electromagnetic field, where one object can affect another object without being in physical contact. For drag to be generated, the solid body must be in contact with the fluid. If there is no fluid, there is no drag. Drag is generated by the difference in velocity between the solid object and the fluid. There must be motion between the object and the fluid. If there is no motion, there is no drag. It makes no difference whether the object moves through a static fluid or whether the fluid moves past a static solid object.

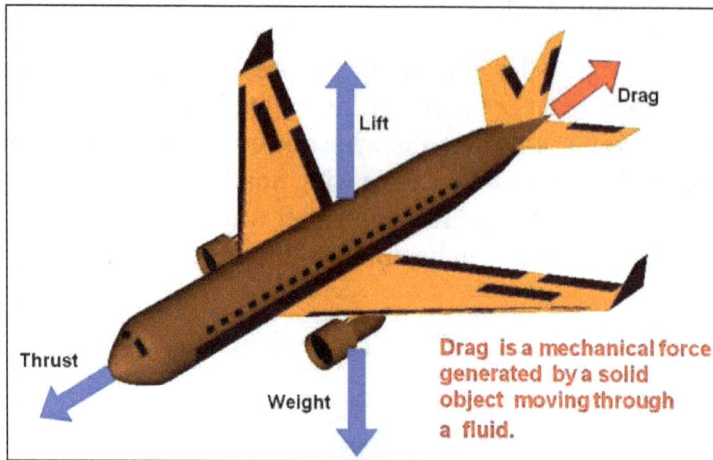

Drag is a force and is therefore a vector quantity having both a magnitude and a direction. Drag acts in a direction that is opposite to the motion of the aircraft. Lift acts perpendicular to the motion. There are many factors that affect the magnitude of the drag. Many of the factors also affect lift but there are some factors that are unique to aircraft drag.

We can think of drag as aerodynamic friction, and one of the sources of drag is the skin friction between the molecules of the air and the solid surface of the aircraft. Because the skin friction is an interaction between a solid and a gas, the magnitude of the skin friction depends on properties of both solid and gas. For the solid, a smooth, waxed surface produces less skin friction than a roughened surface. For the gas, the magnitude depends on the viscosity of the air and the relative magnitude of the viscous forces to the motion of the flow, expressed as the Reynolds number. Along the solid surface, a boundary layer of low energy flow is generated and the magnitude of the skin friction depends on conditions in the boundary layer.

We can also think of drag as aerodynamic resistance to the motion of the object through the fluid. This source of drag depends on the shape of the aircraft and is called form drag. As air flows around a body, the local velocity and pressure are changed. Since pressure is a measure of the momentum of the gas molecules and a change in momentum produces a force, a varying pressure distribution will produce a force on the body. We can determine the magnitude of the force by integrating (or adding up) the local pressure times the surface area around the entire body. The component of the aerodynamic force that is opposed to the motion is the drag; the component perpendicular to the motion is the lift. Both the lift and drag force act through the center of pressure of the object.

There is an additional drag component caused by the generation of lift. Aerodynamicists have named this component the induced drag. It is also called "drag due to lift" because it only occurs on finite, lifting wings. Induced drag occurs because the distribution of lift is not uniform on a wing, but varies from root to tip. For a lifting wing, there is a pressure difference between the upper and lower surfaces of the wing. Vortices are formed at the wing tips, which produce a swirling flow that is very strong near the wing tips and decreases toward the wing root. The local angle of attack of the wing is increased by the induced flow of the tip vortex, giving an additional, downstream-facing, component to the aerodynamic force acting on the wing. The force is called induced drag because it has been "induced" by the action of the tip vortices. The magnitude of induced drag depends on the amount of lift being generated by the wing and on the distribution of lift across the span. Long, thin (chordwise) wings have low induced drag; short wings with a large chord have high induced drag. Wings with an elliptical distribution of lift have the minimum induced drag. Modern airliners use winglets to reduce the induced drag of the wing.

Two additional sources of drag are wave drag and ram drag. As an aircraft approaches the speed of sound, shock waves are generated along the surface. The shock waves produce a change in static pressure and a loss of total pressure. Wave drag is associated with the formation of the shock waves. The magnitude of the wave drag depends on the Mach number of the flow. Ram drag is produced when free stream air is brought inside the aircraft. Jet engines bring air on board, mix the air with fuel; burn the fuel, then exhausts the combustion products to produce thrust. If we look at the basic thrust equation, there is a mass flow time's entrance velocity term that is subtracted from the gross thrust. This "negative thrust" term is the ram drag. Cooling inlets on the aircraft are also sources of ram drag.

Drag Coefficient

The drag coefficient of an object in a moving fluid influence drags force. Any object moving through a fluid experiences drag - the net force in the direction of flow due to pressure and shear stress forces on the surface of the object.

The drag force can be expressed as:

$$F_d = c_d \, 1/2 \, \rho v^2 \, A$$

where,

- F_d = *Drag force (N).*
- c_d = *Drag coefficient.*
- ρ = Density of fluid (*1.2 kg/m³ for air at* NTP).

- v = Flow velocity (m/s).

- A = Characteristic frontal area of the body (m²).

The drag coefficient is a function of several parameters like shape of the body, Reynolds Number for the flow, Froude number, Mach number and Roughness of the Surface.

The characteristic frontal area - A - depends on the body.

Objects drag coefficients are mostly results of experiments. The drag coefficients for some common bodies are indicated below:

Type of Object	Drag Coefficient - cd -	Frontal Area
Laminar flat plate (Re=106)	0.001	
Dolphin	0.0036	wetted area
Turbulent flat plate (Re=106)	0.005	
Subsonic Transport Aircraft	0.012	
Supersonic Fighter, M=2.5	0.016	
Streamline body	0.04	π / 4 d2
Airplane wing, normal position	0.05	
Long stream-lined body	0.1	
Airplane wing, stalled	0.15	
Modern car like Toyota Prius	0.26	frontal area
Sports car, sloping rear	0.2 - 0.3	frontal area
Common car like Opel Vectra (class C)	0.29	frontal area
Hollow semi-sphere facing stream	0.38	
Bird	0.4	frontal area
Solid Hemisphere	0.42	π / 4 d2
Sphere	0.5	
Saloon Car, stepped rear	0.4 - 0.5	frontal area
Convertible, open top	0.6 - 0.7	frontal area
Bus	0.6 - 0.8	frontal area
Old Car like a T-ford	0.7 - 0.9	frontal area
Cube	0.8	s2
Bike racing	0.88	3.9
Bicycle	0.9	
Tractor Trailed Truck	0.96	frontal area
Truck	0.8 - 1.0	frontal area
Person standing	1.0 – 1.3	
Bicycle Upright Commuter	1.1	5.5
Thin Disk	1.1	π / 4 d2
Solid Hemisphere flow normal to flat side	1.17	π / 4 d2

Squared flat plate at 90 deg	1.17	
Wires and cables	1.0 - 1.3	
Person (upright position)	1.0 - 1.3	
Hollow semi-cylinder opposite stream	1.2	
Ski jumper	1.2 - 1.3	
Hollow semi-sphere opposite stream	1.42	
Passenger Train	1.8	frontal area
Motorcycle and rider	1.8	frontal area
Long flat plate at 90 deg	1.98	
Rectangular box	2.1	

Example of Air Resistance Force Acting on a Normal Car

The force required overcoming air resistance for a normal family car with drag coefficient 0.29 and frontal area 2 m² in 90 km/h can be calculated as:

$$Fd = 0.29 \, 1/2 \, (1.2 \text{ kg/m}^3) \, ((90 \text{ km/h}) \, (1000 \text{ m/km})/ \, (3600 \text{ s/h}))^2 \, (2 \text{ m}^2)$$
$$= 217.5 \text{ N}$$

Compare car air resistance with car rolling resistance.

The work done to overcome the air resistance in one hour driving (90 km) can be calculated as:

$$W_d = (217.5 \text{ N}) \, (90 \text{ km}) \, (1000 \text{ m/km})$$

$$= 19575000 \text{ (Nm, J)}$$

The power required to overcome the air resistance when driving 90 km/h can be calculated as:

$$P_d = (217.5 \text{ N}) \, (90 \text{ km/h}) \, (1000 \text{ m/km}) \, (1/3600 \text{ h/s})$$

$$= 5436 \text{ (Nm/s, J/s, W)}$$

$$= 5.4 \text{ (kW)}.$$

Lift

Lift is the force that directly opposes the weight of an airplane and holds the airplane in the air. Lift is generated by every part of the airplane, but most of the lift on a normal airliner is generated by the wings. Lift is a mechanical aerodynamic force produced by the motion of the airplane through the air. Because lift is a force, it is a vector quantity, having both a magnitude and a direction associated with it. Lift acts through the center of pressure of the object and is directed perpendicular to the flow direction. There are several factors which affect the magnitude of lift.

Generation of Lift

There are many explanations for the generation of lift found in encyclopaedias, in basic physics textbooks, and on Web sites. Unfortunately, many of the explanations are misleading and incorrect. Theories on the generation of lift have become a source of great controversy and a topic for heated arguments.

Lift occurs when a moving flow of gas is turned by a solid object. The flow is turned in one direction, and the lift is generated in the opposite direction, according to Newton's Third Law of action and reaction. Because air is a gas and the molecules are free to move about, any solid surface can deflect a flow. For an aircraft wing, both the upper and lower surfaces contribute to the flow turning. Neglecting the upper surface's part in turning the flow leads to an incorrect theory of lift.

No Fluid and No Lift

Lift is a mechanical force. It is generated by the interaction and contact of a solid body with a fluid (liquid or gas). It is not generated by a force field, in the sense of a gravitational field, or an electromagnetic field, where one object can affect another object without being in physical contact. For lift to be generated, the solid body must be in contact with the fluid: no fluid, no lift. The Space Shuttle does not stay in space because of lift from its wings but because of orbital mechanics related to its speed. Space is nearly a vacuum. Without air, there is no lift generated by the wings.

No Motion and No Lift

Lift is generated by the difference in velocity between the solid object and the fluid. There must be motion between the object and the fluid: no motion, no lift. It makes no difference whether the object moves through a static fluid, or the fluid moves past a static solid object. Lift acts perpendicular to the motion. Drag acts in the direction opposed to the motion.

The pilot can control the lift. Any time the control yoke or stick is moved fore or aft, the angle of attack is changed. As the angle of attack increases, lift increases (all other factors being equal). When the aircraft reaches the maximum angle of attack, lift begins to diminish rapidly. This is the stalling angle of attack, known as CLIMAX critical angle of attack. Examine figure, noting how the CL increases until the critical angle of attack is reached, then decreases rapidly with any further increase in the angle of attack.

The shape of the wing or rotor cannot be effective unless it continually keeps "attacking" new air. If an aircraft is to keep flying, the lift-producing airfoil must keep moving. In a helicopter or gyroplane, this is accomplished by the rotation of the rotor blades.

For other types of aircraft, such as airplanes, weight shift control, or gliders, air must be moving across the lifting surface. This is accomplished by the forward speed of the aircraft. Lift is proportional to the square of the aircraft's velocity. For example, an airplane traveling at 200 knots has four times the lift as the same airplane traveling at 100 knots, if the angle of attack and other factors remain constant.

$$L = \frac{C_L \cdot \rho \cdot V^2 \cdot S}{2}$$

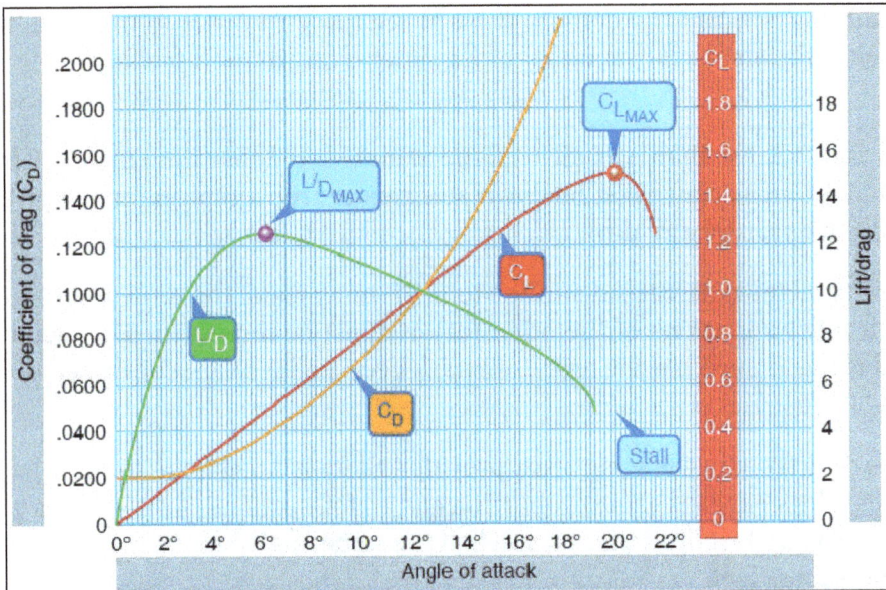

Coefficients of lift and drag at various angles of attack.

The above lift equation exemplifies this mathematically and supports that doubling of the airspeed will result in four times the lift. As a result, one can see that velocity is an important component to the production of lift, which itself can be affected through varying angle of attack. When examining the equation, lift (L) is determined through the relationship of the air density (ρ), the airfoil velocity (V), the surface area of the wing (S) and the coefficient of lift (CL) for a given airfoil. Taking the equation further, one can see an aircraft could not continue to travel in level flight at a constant altitude and maintain the same angle of attack if the velocity is increased. The lift would increase and the aircraft would climb as a result of the increased lift force or speed up.

Therefore, to keep the aircraft straight and level (not accelerating upward) and in a state of equilibrium, as velocity is increased, lift must be kept constant. This is normally accomplished by reducing the angle of attack by lowering the nose. Conversely, as the aircraft is slowed, the decreasing velocity requires increasing the angle of attack to maintain lift sufficient to maintain flight. There is, of course, a limit to how far the angle of attack can be increased, if a stall is to be avoided.

All other factors being constant, for every angle of attack there is a corresponding airspeed required to maintain altitude in steady, uncelebrated flight (true only if maintaining level flight). Since an airfoil always stalls at the same angle of attack, if increasing weight, lift must also be increased. The only method of increasing lift is by increasing velocity if the angle of attack is held constant just short of the "critical," or stalling, angle of attack (assuming no flaps or other high lift devices).

Lift and drag also vary directly with the density of the air. Density is affected by several factors: pressure, temperature, and humidity. At an altitude of 18,000 feet, the density of the air has one-half the density of air at sea level. In order to maintain its lift at a higher altitude, an aircraft must fly at a greater true airspeed for any given angle of attack.

Warm air is less dense than cool air, and moist air is less dense than dry air. Thus, on a hot humid day, an aircraft must be flown at a greater true airspeed for any given angle of attack than on a cool, dry day.

If the density factor is decreased and the total lift must equal the total weight to remain in flight, it follows that one of the other factors must be increased. The factor usually increased is the airspeed or the angle of attack because these are controlled directly by the pilot.

Lift varies directly with the wing area, provided there is no change in the wing's planform. If the wings have the same proportion and airfoil sections, a wing with a planform area of 200 square feet lifts twice as much at the same angle of attack as a wing with an area of 100 square feet.

Two major aerodynamic factors from the pilot's viewpoint are lift and airspeed because they can be controlled readily and accurately. Of course, the pilot can also control density by adjusting the altitude and can control wing area if the aircraft happens to have flaps of the type that enlarge wing area. However, for most situations, the pilot controls lift and airspeed to maneuver an aircraft. For instance, in straight-and-level flight, cruising along at a constant altitude, altitude is maintained by adjusting lift to match the aircraft's velocity or cruise airspeed, while maintaining a state of equilibrium in which lift equals weight. In an approach to landing, when the pilot wishes to land as slowly as practical, it is necessary to increase angle of attack near maximum to maintain lift equal to the weight of the aircraft.

Lift to Drag Ratio

The lift-to-drag ratio (L/D) is the amount of lift generated by a wing or airfoil compared to its drag. A ratio of L/D indicates airfoil efficiency. Aircraft with higher L/D ratios are more efficient than those with lower L/D ratios. In unaccelerated flight with the lift and drag data steady, the proportions of the coefficient of lift (C_L) and coefficient of drag (C_D) can be calculated for specific angle of attack.

The coefficient of lift is dimensionless and relates the lift generated by a lifting body, the dynamic pressure of the fluid flow around the body, and a reference area associated with the body. The coefficient of drag is also dimensionless and is used to quantify the drag of an object in a fluid environment, such as air, and is always associated with a particular surface area.

The L/D ratio is determined by dividing the C_L by the C_D, which is the same as dividing the lift equation by the drag equation as all of the variables, aside from the coefficients, cancel out. The lift and drag equations are as follows (L = Lift in pounds; D = Drag; C_L = coefficient of lift; ρ = density (expressed in slugs per cubic feet); V = velocity (in feet per second); q = dynamic pressure per square foot (q = ½ ρv^2); S = the area of the lifting body (in square feet); and C_D = Ratio of drag pressure to dynamic pressure):

$$D = \frac{C_D \cdot \rho \cdot V^2 \cdot S}{2}$$

Typically at low angle of attack, the coefficient of drag is low and small changes in angle of attack create only slight changes in the coefficient of drag. At high angle of attack, small changes in the angle of attack cause significant changes in drag. The shape of an airfoil, as well as changes in the angle of attack, affects the production of lift.

Notice in figure that the coefficient of lift curve (red) reaches its maximum for this particular wing section at 20° angle of attack and then rapidly decreases. 20° angle of attack is therefore the critical angle of attack. The coefficient of drag curve (orange) increases very rapidly from 14° angle of attack and completely overcomes the lift curve at 21° angle of attack. The lift/drag ratio (green) reaches its maximum at 6° angle of attack, meaning that at this angle, the most lift is obtained for the least amount of drag.

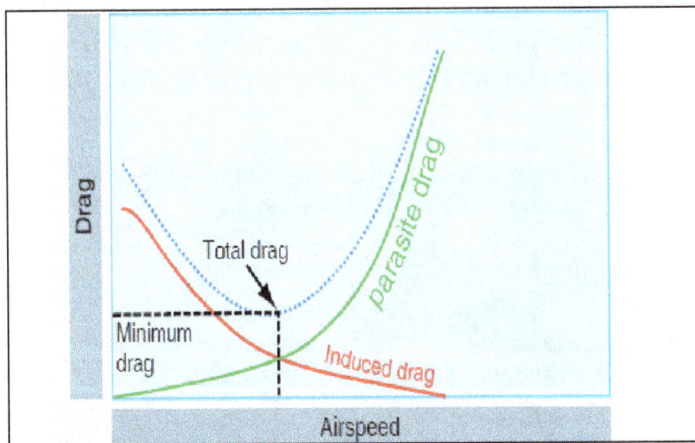

Drag versus speed.

Note that the maximum lift/drag ratio (L/D_{MAX}) occurs at one specific C_L and angle of attack. If the aircraft is operated in steady flight at L/D_{MAX}, the total drag is at a minimum. Any angle of attack lower or higher than that for L/D_{MAX} reduces the L/D and consequently increases the total drag for a given aircraft's lift. Figure depicts the L/D_{MAX} by the lowest portion of the blue line labeled "total drag." The configuration of an aircraft has a great effect on the L/D.

Thrust

Thrust is the force which moves an aircraft through the air. Thrust is used to overcome the drag of an airplane, and to overcome the weight of a rocket. Thrust is generated by the engines of the aircraft through some kind of propulsion system.

Thrust is a mechanical force, so the propulsion system must be in physical contact with a working fluid to produce thrust. Thrust is generated most often through the reaction of accelerating a mass of gas. Since thrust is a force, it is a vector quantity having both a magnitude and a direction. The engine does work on the gas and accelerates the gas to the rear of the engine; the thrust is generated in the opposite direction from the accelerated gas. The magnitude of the thrust depends on the amount of gas that is accelerated and on the difference in velocity of the gas through the engine.

Thrust Reversal

In a typical turbofan jet engine the oncoming airflow is compressed throughout a series of compressor stages, mixed with a fuel (typically kerosene) and combusted, drastically increasing pressure and temperature, and then expanded through a nozzle to provide thrust towards the rear of the aircraft. By accelerating the fluid towards aft, Newton's Third Law implies that this impulse must be reacted by an equal and opposite force in the opposite direction, thus propelling the aircraft forward. However, modern jet engines are also capable of producing thrust in the opposing direction.

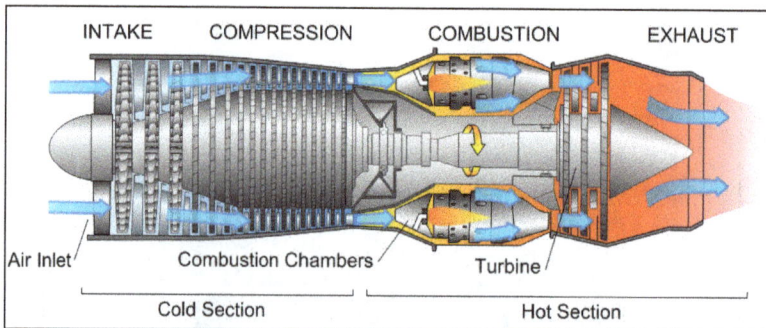

Diagram of a typical gas turbine jet engine.

Air is compressed by the fan blades as it enters the engine, and it is mixed and burned with fuel in the combustion section. The hot exhaust gases provide forward thrust and turn the turbines which drive the compressor fan blades.

Thrust reversal is achieved by momentarily diverting the hot exhaust gases towards the front of the aircraft or changing the propeller/compressor pitch so that the thrust produced is directed forward. Thus thrust will act against the forward direction of travel and provide a means of deceleration. Thrust reversal is used in some flight scenarios in order to:

- Alleviate the stress and reduce wear on the brakes or to enable shorter landing distances. Reverse thrust can reduce the braking distance by a third or more.

- Momentarily increase the braking force during emergencies or just after touchdown when the aircraft is still traveling at a high velocity and the residual aerodynamic lift is significant. Lift reduces the normal reaction force with the ground and therefore limits friction and grip on the tyres.

- Rapid deceleration in flight to enable quick changes of speed. Most aircraft cannot operate thrust reversal in flight and the majority that can are propeller-driven.

- Helping to push an aircraft back from a gate. A maneuver called "powerback".

Almost everyone who has sat in a row near the wings will have heard reverse thrust in action before. Next time you land wait for the sudden high-pitched increase in engine noise just after touchdown.

The method to achieve thrust reversal varies greatly between the different types of engines:

- Since the 1930s propeller-driven aircraft generate reverse thrust by changing the angle of attack of their controllable pitch propellers:

  ◦ Older reciprocating engines and modern turboprop engines both have the ability to set the propeller angle to "flat pitch". As a result the propellor airfoils produce no forward or reverse thrust, but large amounts of drag instead. This allows the engine speed to be kept at a constant speed while descending.

  ◦ The classic approach is to pitch the propeller blades to a negative angle of attack in order to direct the thrust forward.

- In jet engines thrust reversal is not accomplished by running the engine in reverse but by diverting the high-velocity exhaust jet blast to the front of the engines. This can be achieved in different ways:

  ◦ The target-type thrust reverser: After the combustion chamber, reverser blades angle outward in order to prematurely redirect the high-speed jet radially outwards and towards the front of the engine. This construction generally gives the appearance of flower petals.

  ◦ The clamshell type: Two reverser buckets are hinged at the aft of the engine, and when deployed, intrude into the exhaust of the engine. In this manner the jet blast is captured and re-oriented towards the front.

○ In a turbofan engine some of the air intake is not passed through the main part of the engine, but redirected along an outside channel without being combusted. This bypass duct is aptly named "cold flow" and this arrangement is used to save fuel and reduce engine noise. Furthermore, the bypass flow can also be used to channel air radially outwards and forwards to provide thrust reversal.

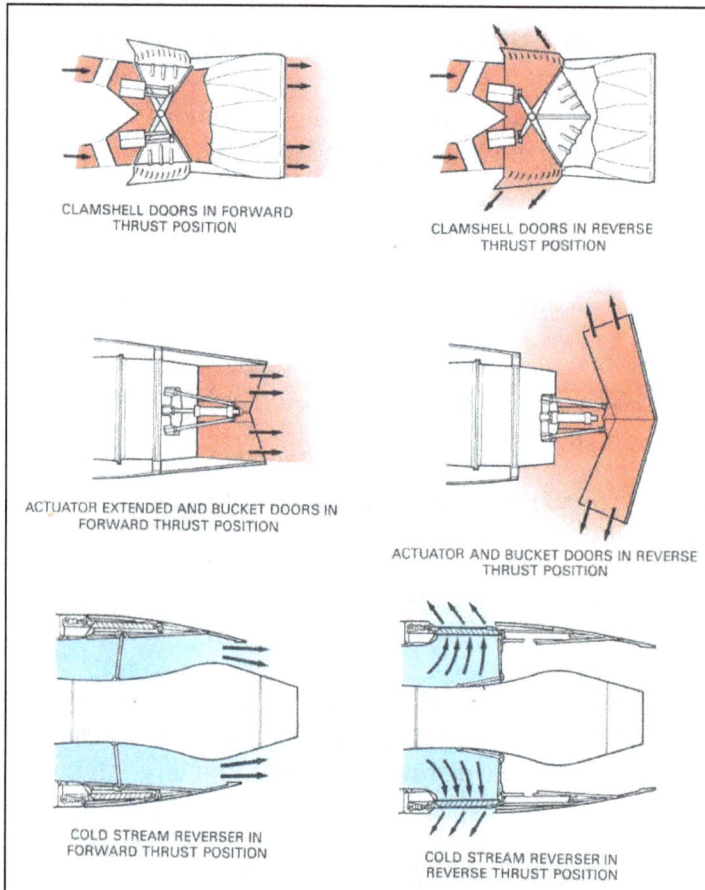

CLAMSHELL DOORS IN FORWARD THRUST POSITION

CLAMSHELL DOORS IN REVERSE THRUST POSITION

ACTUATOR EXTENDED AND BUCKET DOORS IN FORWARD THRUST POSITION

ACTUATOR AND BUCKET DOORS IN REVERSE THRUST POSITION

COLD STREAM REVERSER IN FORWARD THRUST POSITION

COLD STREAM REVERSER IN REVERSE THRUST POSITION

The three different types of thrust mechanisms.

Helicopter Aerodynamics

Once a helicopter leaves the ground, it is acted upon by the four aerodynamic forces.

Powered Flight

In powered flight (hovering, vertical, forward, sideward, or rearward), the total lift and thrust forces of a rotor are perpendicular to the tip-path plane or plane of rotation of the rotor.

Hovering Flight

During hovering flight, a helicopter maintains a constant position over a selected point, usually a few feet above the ground. For a helicopter to hover, the lift and thrust produced by the rotor system act straight up and must equal the weight and drag, which act straight down. While hovering, you can change the amount of main rotor thrust to maintain the desired hovering altitude. This is done by changing the angle of attack of the main rotor blades and by varying power, as needed. In this case, thrust acts in the same vertical direction as lift.

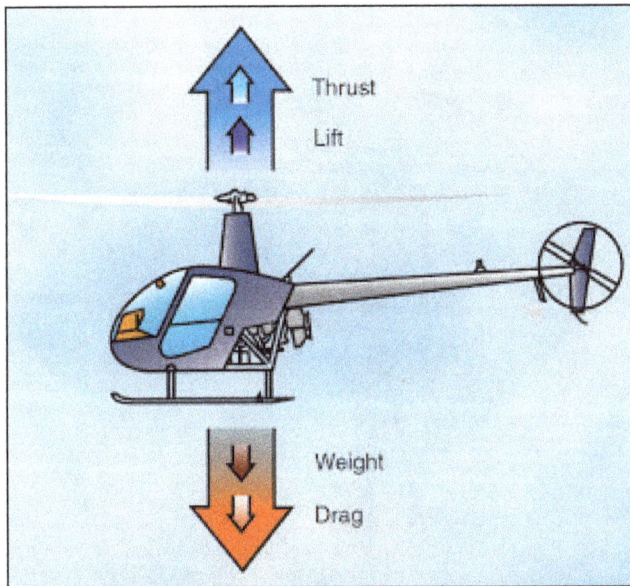

Thrust

Lift

Weight

Drag

To maintain a hover at a constant altitude, enough lift and thrust must be generated to equal the weight of the helicopter and the drag produced by the rotor blades.

The weight that must be supported is the total weight of the helicopter and its occupants. If the amount of thrust is greater than the actual weight, the helicopter gains altitude; if thrust is less than weight, the helicopter loses altitude.

The drag of a hovering helicopter is mainly induced drag incurred while the blades are producing lift. There is, however, some profile drag on the blades as they rotate through the air.

An important consequence of producing thrust is torque. For every action there is an equal and opposite reaction. Therefore, as the engine turns the main rotor system in a counter clockwise direction, the helicopter fuselage turns clockwise. The amount of torque is directly related to the amount of engine power being used to turn the main rotor system. Remember, as power changes, torque changes.

To counteract this torque-induced turning tendency, an antitorque rotor or tail rotor is incorporated into most helicopter designs. You can vary the amount of thrust produced by the tail rotor in relation to the amount of torque produced by the engine. As the engine supplies more power, the tail rotor must produce more thrust. This is done through the use of antitorque pedals.

Translating Tendency or Drift

During hovering flight, a single main rotor helicopter tends to drift in the same direction as antitorque rotor thrust. This drifting tendency is called translating tendency.

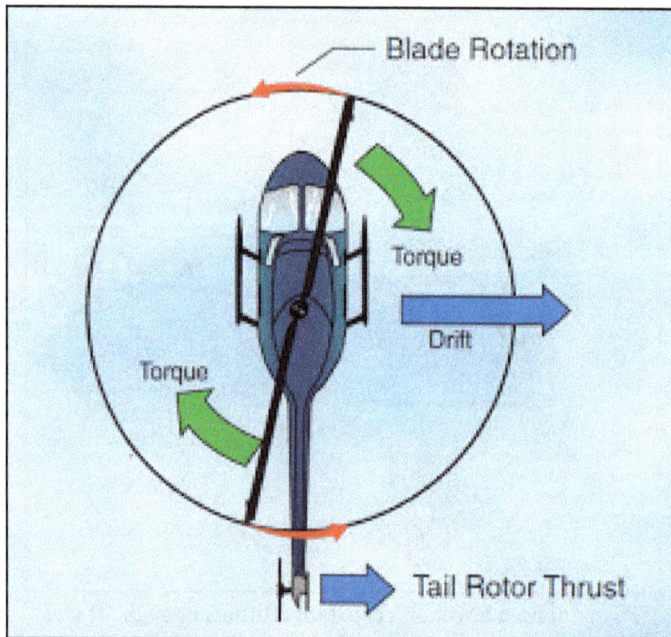

A tail rotor is designed to produce thrust in a direction opposite torque.
The thrust produced by the tail rotor is sufficient to move the helicopter laterally.

To counteract this drift, one or more of the following features may be used:

- The main transmission is mounted so that the rotor mast is rigged for the tip-path plane to have a builtin tilt opposite tail thrust, thus producing a small sideward thrust.

- Flight control rigging is designed so that the rotor disc is tilted slightly opposite tail rotor thrust when the cyclic is centered.

- The cyclic pitch control system is designed so that the rotor disc tilts slightly opposite tail rotor thrust when in a hover.

Counteracting translating tendency, in a helicopter with a counterclockwise main rotor system, causes the left skid to hang lower while hovering. The opposite is true for rotor systems turning clockwise when viewed from above.

Pendular Action

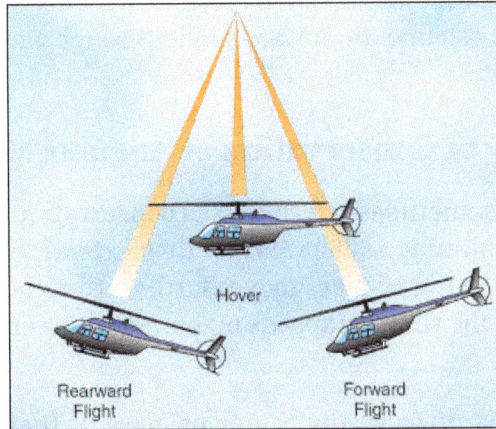

Because the helicopter's body has mass and is suspended from a
single point (the rotor mast head), it tends to act much like a pendulum.

Since the fuselage of the helicopter, with a single main rotor, is suspended from a single point and has considerable mass, it is free to oscillate either longitudinally or laterally in the same way as a pendulum. This pendular action can be exaggerated by over controlling; therefore, control movements should be smooth and not exaggerated.

Coning

In order for a helicopter to generate lift, the rotor blades must be turning. This creates a relative wind that is opposite the direction of rotor system rotation. The rotation of the rotor system creates centrifugal force (inertia), which tends to pull the blades straight outward from the main rotor hub. The faster the rotation, the greater the centrifugal force. This force gives the rotor blades their rigidity and, in turn, the strength to support the weight of the helicopter. The centrifugal force generated determines the maximum operating rotor r.p.m. due to structural limitations on the main rotor system.

As a vertical takeoff is made, two major forces are acting at the same time—centrifugal force acting outward and perpendicular to the rotor mast, and lift acting upward and parallel to the mast. The result of these two forces is that the blades assume a conical path instead of remaining in the plane perpendicular to the mast.

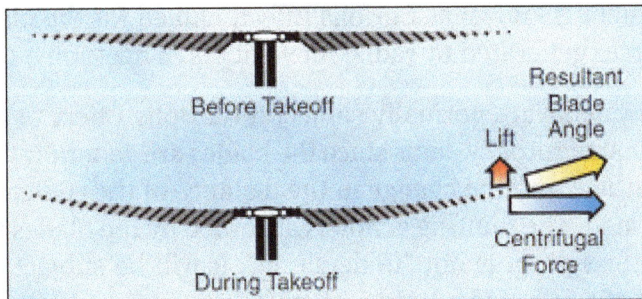

Rotor blade coning occurs as the rotor blades begin to lift the weight of the helicopter. In a semirigid and rigid rotor system, coning results in blade bending. In an articulated rotor system, the blades assume an upward angle through movement about the flapping hinges.

Coriolis Effect: Law of Conservation of Angular Momentum

Coriolis Effect, which is sometimes referred to as conservation of angular momentum, might be compared to spinning skaters. When they extend their arms, their rotation slows down because the center of mass moves farther from the axis of rotation. When their arms are retracted, the rotation speeds up because the center of mass moves closer to the axis of rotation.

When a rotor blade flaps upward, the center of mass of that blade moves closer to the axis of rotation and blade acceleration takes place in order to conserve angular momentum. Conversely, when that blade flaps downward, its center of mass moves further from the axis of rotation and blade deceleration takes place. Keep in mind that due to coning, a rotor blade will not flap below a plane passing through the rotor hub and perpendicular to the axis of rotation. The acceleration and deceleration actions of the rotor blades are absorbed by either dampers or the blade structure itself, depending upon the design of the rotor system.

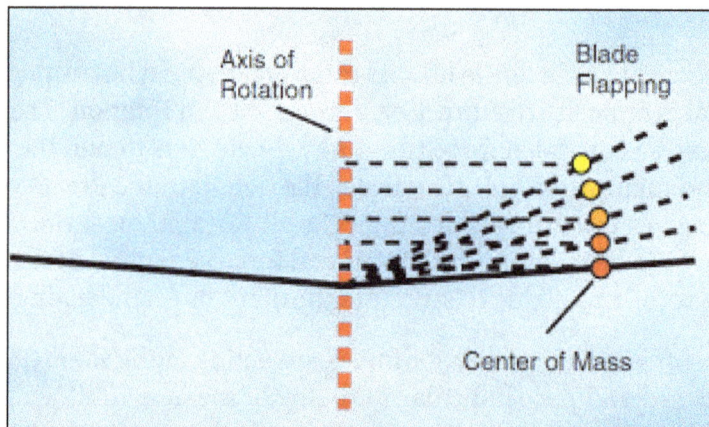

The tendency of a rotor blade to increase or decrease its velocity in its plane of rotation due to mass movement is known as Coriolis Effect, named for the mathematician who made studies of forces generated by radial movements of mass on a rotating disc.

Two-bladed rotor systems are normally subject to Coriolis Effect to a much lesser degree than are articulated rotor systems since the blades are generally "underslung" with respect to the rotor hub, and the change in the distance of the center of mass from the axis of rotation is small. The hunting action is absorbed by the blades through bending. If a two-bladed rotor system is not "underslung," it will be subject to Coriolis Effect comparable to that of a fully articulated system.

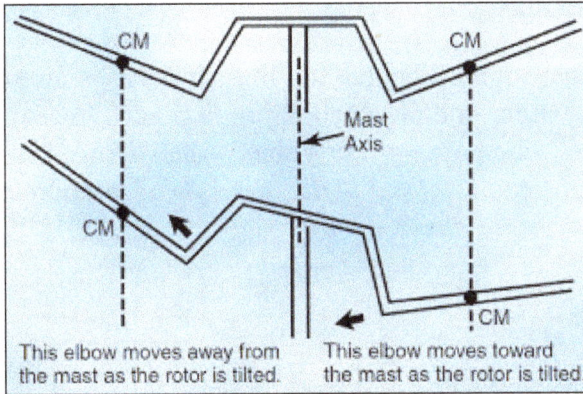

CM · CM · Mast Axis · CM · CM

This elbow moves away from the mast as the rotor is tilted.

This elbow moves toward the mast as the rotor is tilted.

Because of the underslung rotor, the center of mass remains approximately the same distance from the mast after the rotor is tilted.

Ground Effect

When hovering near the ground, a phenomenon known as ground effect takes place. This effect usually occurs less than one rotor diameter above the surface. As the induced airflow through the rotor disc is reduced by the surface friction, the lift vector increases. This allows a lower rotor blade angle for the same amount of lift, which reduces induced drag. Ground effect also restricts the generation of blade tip vortices due to the downward and outward airflow making a larger portion of the blade produce lift. When the helicopter gains altitude vertically, with no forward airspeed, induced airflow is no longer restricted, and the blade tip vortices increase with the decrease in outward airflow. As a result, drag increases which means a higher pitch angle, and more power is needed to move the air down through the rotor.

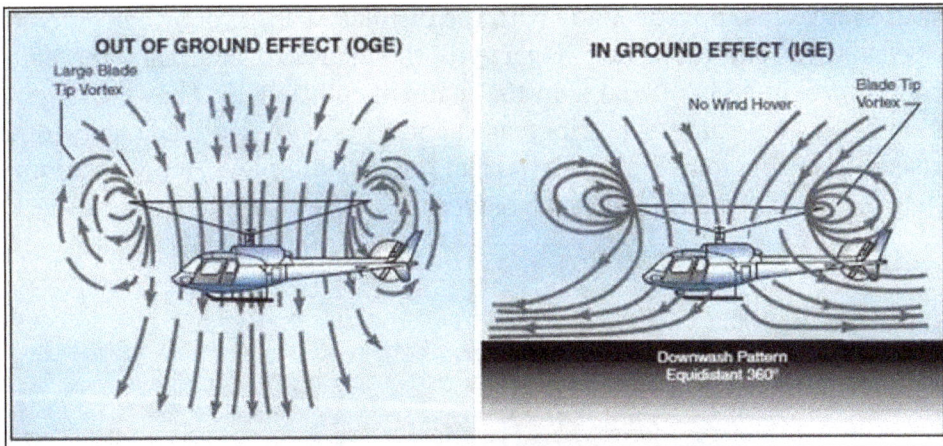

OUT OF GROUND EFFECT (OGE)
Large Blade Tip Vortex

IN GROUND EFFECT (IGE)
No Wind Hover
Blade Tip Vortex

Downwash Pattern Equidistant 360°

Air circulation patterns change when hovering out of ground effect (OGE) and when hovering in ground effect (IGE).

Ground effect is at its maximum in a no-wind condition over a firm, smooth surface. Tall grass, rough terrain, revetments, and water surfaces alter the airflow pattern, causing an increase in rotor tip vortices.

Gyroscopic Precession

The spinning main rotor of a helicopter acts like a gyroscope. As such, it has the properties of gyroscopic action, one of which is precession. Gyroscopic precession is the resultant action or deflection of a spinning object when a force is applied to this object. This action occurs approximately 90° in the direction of rotation from the point where the force is applied.

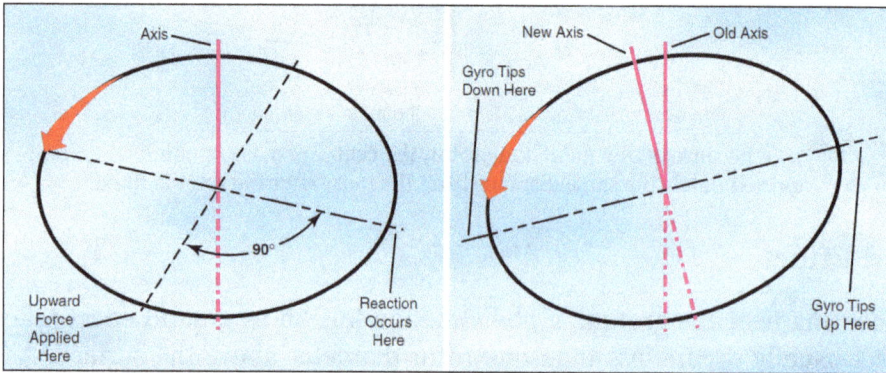

Gyroscopic precession principle—when a force is applied to a spinning gyro, the maximum reaction occurs approximately 90° later in the direction of rotation.

Let us look at a two-bladed rotor system to see how gyroscopic precession affects the movement of the tippath plane. Moving the cyclic pitch control increases the angle of attack of one rotor blade with the result that a greater lifting force is applied at that point in the plane of rotation. This same control movement simultaneously decreases the angle of attack of the other blade the same amount, thus decreasing the lifting force applied at that point in the plane of rotation. The blade with the increased angle of attack tends to flap up; the blade with the decreased angle of attack tends to flap down. Because the rotor disk acts like a gyro, the blades reach maximum deflection at a point approximately 90° later in the plane of rotation. As shown in figure, the retreating blade angle of attack is increased and the advancing blade angle of attack is decreased resulting in a tipping forward of the tip-path plane, since maximum deflection takes place 90° later when the blades are at the rear and front, respectively.

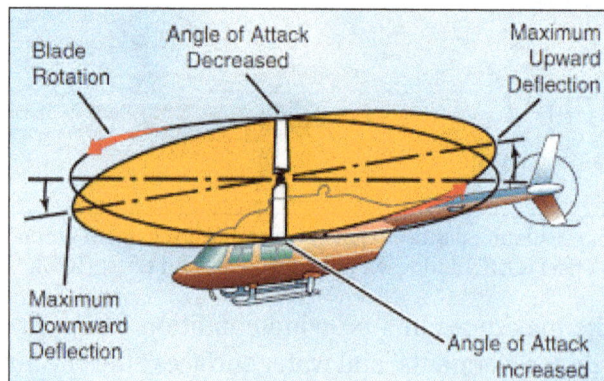

With a counterclockwise main rotor blade rotation, as each blade passes the 90° position on the left, the maximum increase in angle of attack occurs. As each blade passes the 90° position to the right, the maximum decrease in angle of attack occurs. Maximum deflection takes place 90° later—maximum upward deflection at the rear and maximum downward deflection at the front—and the tip-path plane tips forward.

In a rotor system using three or more blades, the movement of the cyclic pitch control changes the angle of attack of each blade an appropriate amount so that the end result is the same.

Vertical Flight

Hovering is actually an element of vertical flight. Increasing the angle of attack of the rotor blades (pitch) while their velocity remains constant generates additional vertical lift and thrust and the helicopter ascends. Decreasing the pitch causes the helicopter to descend. In a no wind condition when lift and thrust are less than weight and drag, the helicopter descends vertically. If lift and thrust are greater than weight and drag, the helicopter ascends vertically.

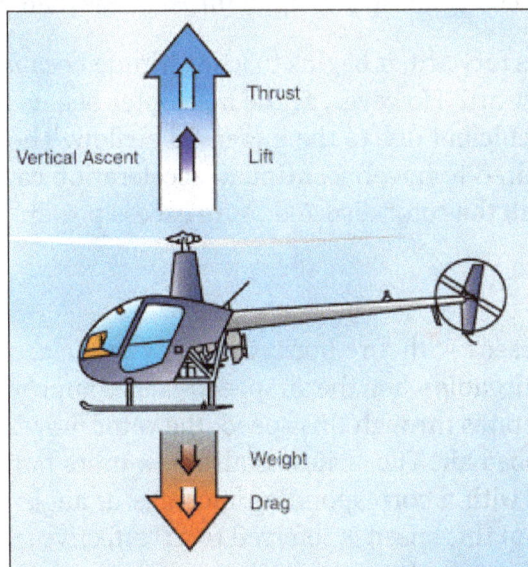

To ascend vertically, more lift and thrust must be generated
to overcome the forces of weight and the drag.

Forward Flight

In or during forward flight, the tip-path plane is tilted forward, thus tilting the total lift-thrust force forward from the vertical. This resultant lift-thrust force can be resolved into two components—lift acting vertically upward and thrust acting horizontally in the direction of flight. In addition to lift and thrust, there is weight (the downward acting force) and drag (the rearward acting or retarding force of inertia and wind resistance).

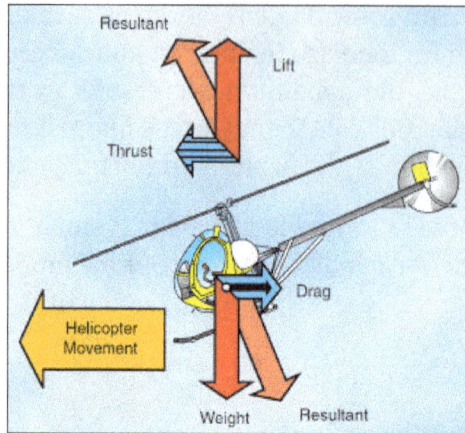

To transition into forward flight, some of the vertical thrust must be vectored horizontally. You initiate this by forward movement of the cyclic control.

In straight-and-level, unaccelerated forward flight, lift equals weight and thrust equals drag (straight-and-level flight is flight with a constant heading and at a constant altitude). If lift exceeds weight, the helicopter climbs; if lift is less than weight, the helicopter descends. If thrust exceeds drag, the helicopter speeds up; if thrust is less than drag, it slows down.

As the helicopter moves forward, it begins to lose altitude because of the lift that is lost as thrust is diverted forward. However, as the helicopter begins to accelerate, the rotor system becomes more efficient due to the increased airflow. The result is excess power over that which is required to hover. Continued acceleration causes an even larger increase in airflow through the rotor disc and more excess power.

Translational Lift

Translational lift is present with any horizontal flow of air across the rotor. This increased flow is most noticeable when the airspeed reaches approximately 16 to 24 knots. As the helicopter accelerates through this speed, the rotor moves out of its vortices and is in relatively undisturbed air. The airflow is also now more horizontal, which reduces induced flow and drag with a corresponding increase in angle of attack and lift. The additional lift available at this speed is referred to as "effective translational lift" (ETL).

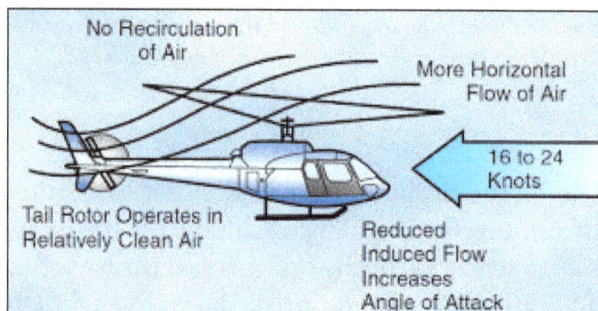

Effective translational lift is easily recognized in actual flight by a transient induced aerodynamic vibration and increased performance of the helicopter.

When a single-rotor helicopter flies through translational lift, the air flowing through the main rotor and over the tail rotor becomes less turbulent and more aerodynamically efficient. As the tail rotor efficiency improves, more thrust is produced causing the aircraft to yaw left in a counterclockwise rotor system. It will be necessary to use right torque pedal to correct for this tendency on takeoff. Also, if no corrections are made, the nose rises or pitches up, and rolls to the right. This is caused by combined effects of dissymmetry of lift and transverse flow effect, and is corrected with cyclic control.

Translational lift is also present in a stationary hover if the wind speed is approximately 16 to 24 knots. In normal operations, always utilize the benefit of translational lift, especially if maximum performance is needed.

Induced Flow

As the rotor blades rotate they generate what is called rotational relative wind. This airflow is characterized as flowing parallel and opposite the rotor's plane of rotation and striking perpendicular to the rotor blade's leading edge. This rotational relative wind is used to generate lift. As rotor blades produce lift, air is accelerated over the foil and projected downward. Anytime a helicopter is producing lift, it moves large masses of air vertically and down through the rotor system. This downwash or induced flow can significantly change the efficiency of the rotor system. Rotational relative wind combines with induced flow to form the resultant relative wind. As induced flow increases, resultant relative wind becomes less horizontal. Since angle of attack is determined by measuring the difference between the chord line and the resultant relative wind, as the resultant relative wind becomes less horizontal, angle of attack decreases.

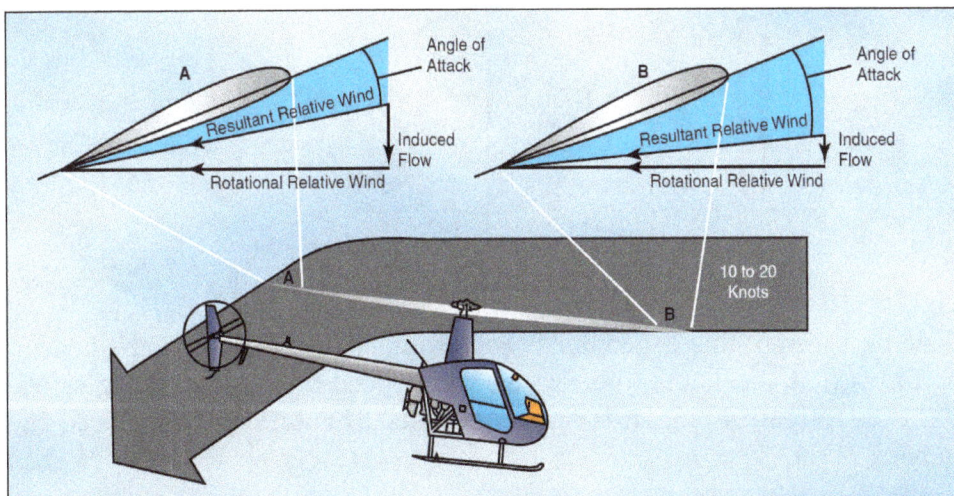

A helicopter in forward flight, or hovering with a headwind or crosswind, has more molecules of air entering the aft portion of the rotor blade. Therefore, the angle of attack is less and the induced flow is greater at the rear of the rotor disc.

Transverse Flow Effect

As the helicopter accelerates in forward flight, induced flow drops to near zero at the forward disc area and increases at the aft disc area. This increases the angle of attack at the front disc area causing the rotor blade to flap up, and reduces angle of attack at the aft disc area causing the rotor blade to flap down. Because the rotor acts like a gyro, maximum displacement occurs 90° in the direction of rotation. The result is a tendency for the helicopter to roll slightly to the right as it accelerates through approximately 20 knots or if the headwind is approximately 20 knots.

You can recognize transverse flow effect because of increased vibrations of the helicopter at airspeeds just below effective translational lift on takeoff and after passing through effective translational lift during landing. To counteract transverse flow effect, a cyclic input needs to be made.

Dissymmetry of Lift

When the helicopter moves through the air, the relative airflow through the main rotor disc is different on the advancing side than on the retreating side. The relative wind encountered by the advancing blade is increased by the forward speed of the helicopter, while the relative wind speed acting on the retreating blade is reduced by the helicopter's forward airspeed. Therefore, as a result of the relative wind speed, the advancing blade side of the rotor disc produces more lift than the retreating blade side. This situation is defined as dissymmetry of lift.

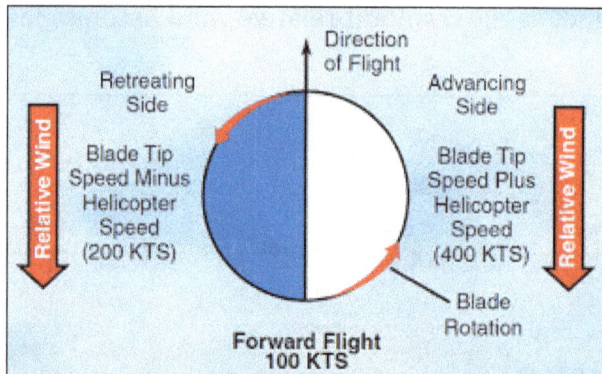

The blade tip speed of this helicopter is approximately 300 knots. If the helicopter is moving forward at 100 knots, the relative wind speed on the advancing side is 400 knots. On the retreating side, it is only 200 knots. This difference in speed causes a dissymmetry of lift.

If this condition was allowed to exist, a helicopter with a counterclockwise main rotor blade rotation would roll to the left because of the difference in lift. In reality, the main rotor blades flap and feather automatically to equalize lift across the rotor disc. Articulated rotor systems, usually with three or more blades, incorporate a horizontal hinge (flapping hinge) to allow the individual rotor blades to move, or flap up and down as they rotate. A semirigid rotor system (two blades) utilizes a teetering hinge, which allows the blades to flap as a unit. When one blade flaps up, the other flaps down.

As shown in figure, as the rotor blade reaches the advancing side of the rotor disc (A), it reaches its maximum upflap velocity. When the blade flaps upward, the angle between the chord line and the resultant relative wind decreases. This decreases the angle of attack, which reduces the amount of lift produced by the blade. At position (C) the rotor blade is now at its maximum downflapping velocity. Due to downflapping, the angle between the chord line and the resultant relative wind increases. This increases the angle of attack and thus the amount of lift produced by the blade.

The combination of blade flapping and slow relative wind acting on the retreating blade normally limits the maximum forward speed of a helicopter. At a high forward speed, the retreating blade stalls because of a high angle of attack and slow relative wind speed. This situation is called retreating blade stall and is evidenced by a nose pitch up, vibration, and a rolling tendency—usually to the left in helicopters with counterclockwise blade rotation.

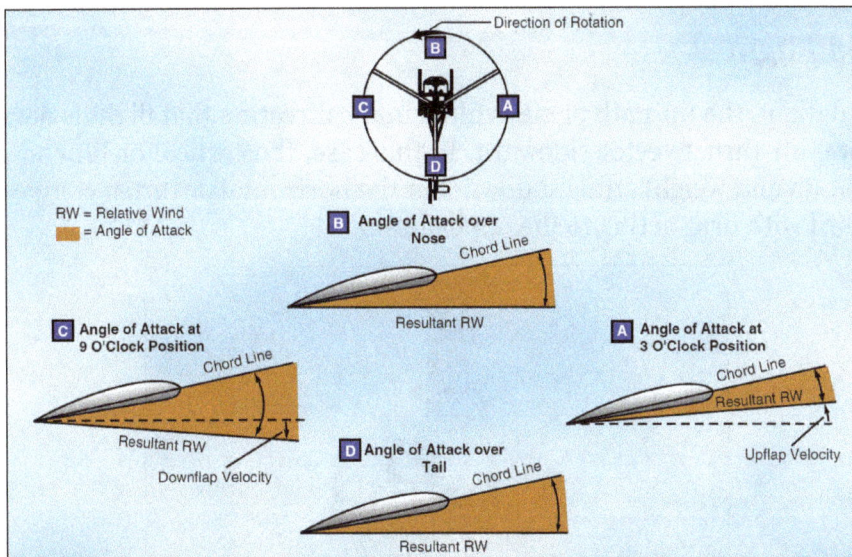

The combined upward flapping (reduced lift) of the advancing blade and downward flapping (increased lift) of the retreating blade equalizes lift across the main rotor disc counteracting dissymmetry of lift.

You can avoid retreating blade stall by not exceeding the never-exceed speed. This speed is designated VNE and is usually indicated on a placard and marked on the airspeed indicator by a red line.

During aerodynamic flapping of the rotor blades as they compensate for dissymmetry of lift, the advancing blade achieves maximum upflapping displacement over the nose and maximum downflapping displacement over the tail. This causes the tip-path plane to tilt to the rear and is referred to as blowback. Figure shows how the rotor disc was originally oriented with the front down following the initial cyclic input, but as airspeed is gained and flapping eliminates dissymmetry of lift, the front of the disc comes up, and the back of the disc goes down. This reorientation of the rotor disc changes the direction in which total rotor thrust acts so that the helicopter's forward speed slows, but can be corrected with cyclic input.

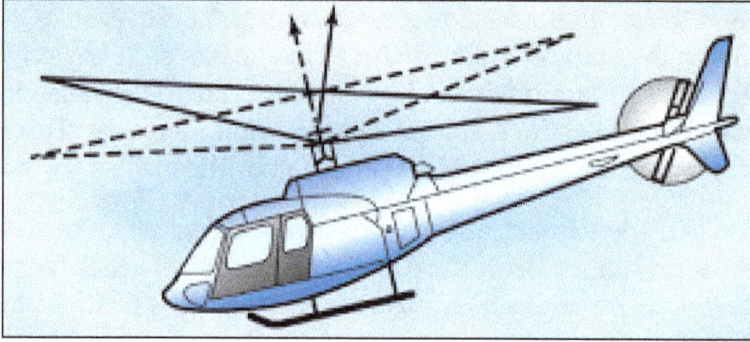

To compensate for blowback, you must move the cyclic forward.
Blowback is more pronounced with higher airspeeds.

Sideward Flight

In sideward flight, the tip-path plane is tilted in the direction that flight is desired. This tilts the total lift-thrust vector sideward. In this case, the vertical or lift component is still straight up and weight straight down, but the horizontal or thrust component now acts sideward with drag acting to the opposite side.

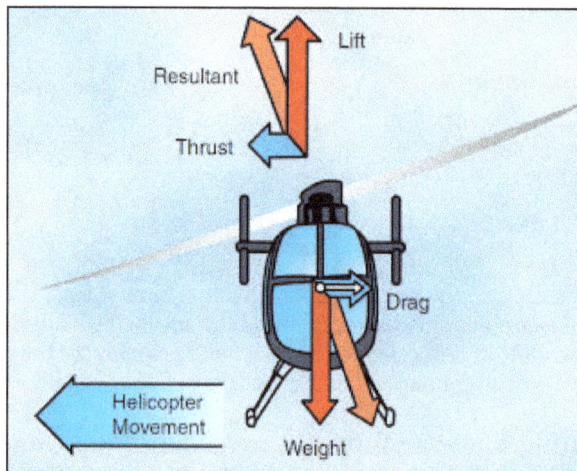

Forces acting on the helicopter during sideward flight.

Rearward Flight

For rearward flight, the tip-path plane is tilted rearward, which, in turn, tilts the lift-thrust vector rearward. Drag now acts forward with the lift component straight up and weight straight down.

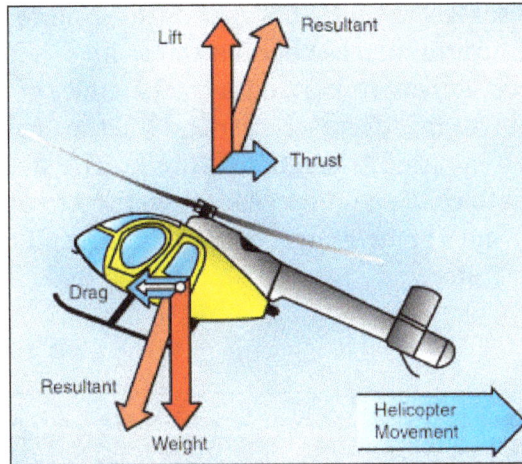

Forces acting on the helicopter during rearward flight.

Turning Flight

In forward flight, the rotor disc is tilted forward, which also tilts the total lift-thrust force of the rotor disc forward. When the helicopter is banked, the rotor disc is tilted sideward resulting in lift being separated into two components. Lift acting upward and opposing weight is called the vertical component of lift. Lift acting horizontally and opposing inertia (centrifugal force) is the horizontal component of lift (centripetal force).

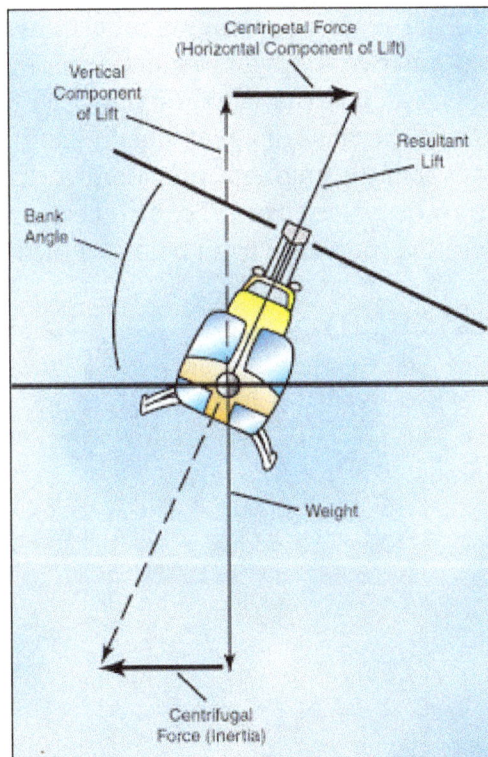

The horizontal component of lift accelerates the helicopter toward the center of the turn.

As the angle of bank increases, the total lift force is tilted more toward the horizontal, thus causing the rate of turn to increase because more lift is acting horizontally. Since the resultant lifting force acts more horizontally, the effect of lift acting vertically is deceased. To compensate for this decreased vertical lift, the angle of attack of the rotor blades must be increased in order to maintain altitude. The steeper the angle of bank, the greater the angle of attack of the rotor blades required to maintain altitude. Thus, with an increase in bank and a greater angle of attack, the resultant lifting force increases and the rate of turn is faster.

Autorotation

During an autorotation, the upward flow of relative wind permits the main rotor blades to rotate at their normal speed. In effect, the blades are "gliding" in their rotational plane.

Autorotation is the state of flight where the main rotor system is being turned by the action of relative wind rather than engine power. It is the means by which a helicopter can be landed safely in the event of an engine failure. In this case, you are using altitude as potential energy and converting it to kinetic energy during the descent and touchdown. All helicopters must have this capability in order to be certified. Autorotation is permitted mechanically because of a freewheeling unit, which allows the main rotor to continue turning even if the engine is not running. In normal powered flight, air is drawn into the main rotor system from above and exhausted downward. During autorotation, airflow enters the rotor disc from below as the helicopter descends.

Autorotation (Vertical Flight)

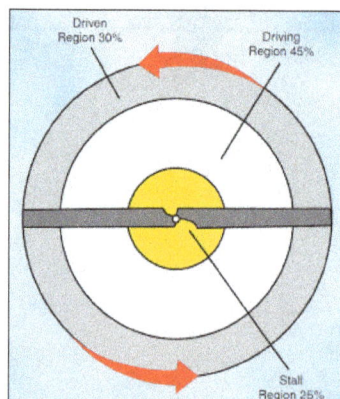

Blade regions in vertical autorotation descent.

Most autorotations are performed with forward speed. For simplicity, the following aerodynamic explanation is based on a vertical autorotative descent (no forward speed) in still air. Under these conditions, the forces that cause the blades to turn are similar for all blades regardless of their position in the plane of rotation. Therefore, dissymmetry of lift resulting from helicopter airspeed is not a factor.

During vertical autorotation, the rotor disc is divided into three regions as illustrated in figure the driven region, the driving region, and the stall region. Figure shows four blade sections that illustrate force vectors. Part A is the driven region, B and D are points of equilibrium, part C is the driving region, and part E is the stall region. Force vectors are different in each region because rotational relative wind is slower near the blade root and increases continually toward the blade tip. Also, blade twist gives a more positive angle of attack in the driving region than in the driven region. The combination of the inflow up through the rotor with rotational relative wind produces different combinations of aerodynamic force at every point along the blade.

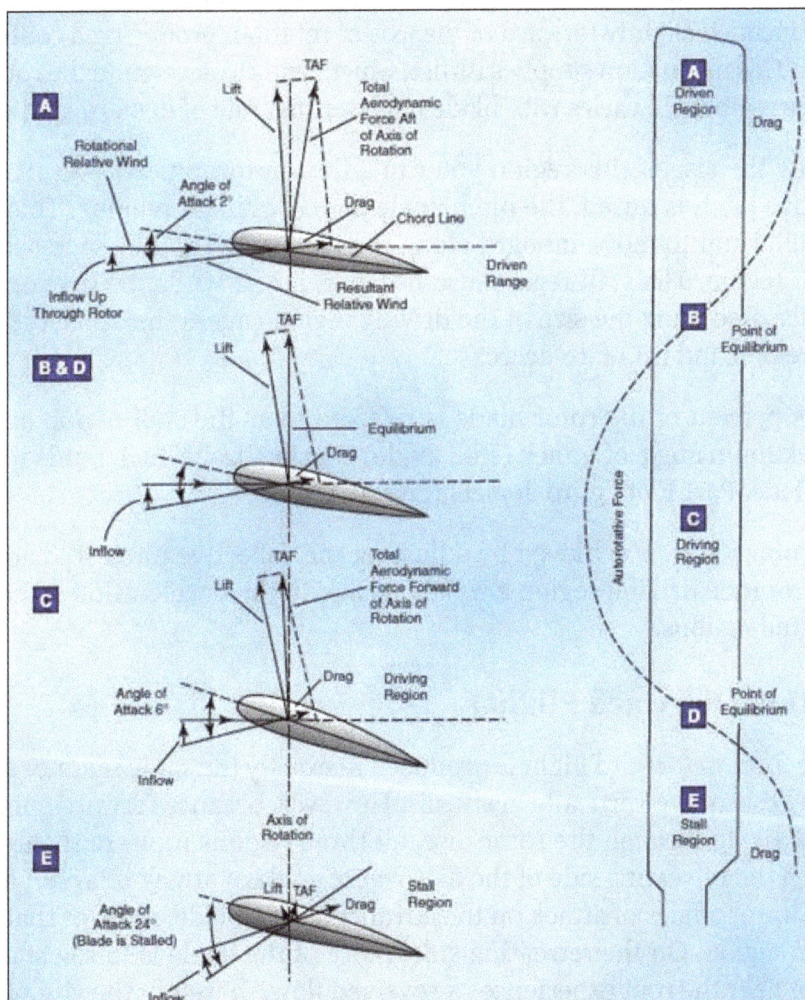

Force vectors in vertical autorotation descent.

The driven region, also called the propeller region, is nearest the blade tips. Normally, it consists of about 30 percent of the radius. In the driven region, part A of figure, the total aerodynamic force acts behind the axis of rotation, resulting in a overall drag force. The driven region produces some lift, but that lift is offset by drag. The overall result is a deceleration in the rotation of the blade. The size of this region varies with the blade pitch, rate of descent, and rotor r.p.m. When changing autorotative r.p.m., blade pitch, or rate of descent, the size of the driven region in relation to the other regions also changes.

There are two points of equilibrium on the blade—one between the driven region and the driving region, and one between the driving region and the stall region. At points of equilibrium, total aerodynamic force is aligned with the axis of rotation. Lift and drag are produced, but the total effect produces neither acceleration nor deceleration.

The driving region, or autorotative region, normally lies between 25 to 70 percent of the blade radius. Part C of figure shows the driving region of the blade, which produces the forces needed to turn the blades during autorotation. Total aerodynamic force in the driving region is inclined slightly forward of the axis of rotation, producing a continual acceleration force. This inclination supplies thrust, which tends to accelerate the rotation of the blade. Driving region size varies with blade pitch setting, rate of descent, and rotor r.p.m.

By controlling the size of this region you can adjust autorotative r.p.m. For example, if the collective pitch is raised, the pitch angle increases in all regions. This causes the point of equilibrium to move inboard along the blade's span, thus increasing the size of the driven region. The stall region also becomes larger while the driving region becomes smaller. Reducing the size of the driving region causes the acceleration force of the driving region and r.p.m. to decrease.

The inner 25 percent of the rotor blade is referred to as the stall region and operates above its maximum angle of attack (stall angle) causing drag which tends to slow rotation of the blade. Part E of figure depicts the stall region.

A constant rotor r.p.m. is achieved by adjusting the collective pitch so blade acceleration forces from the driving region are balanced with the deceleration forces from the driven and stall regions.

Autorotation (Forward Flight)

Autorotative force in forward flight is produced in exactly the same manner as when the helicopter is descending vertically in still air. However, because forward speed changes the inflow of air up through the rotor disc, all three regions move outboard along the blade span on the retreating side of the disc where angle of attack is larger, as shown in figure. With lower angles of attack on the advancing side blade, more of that blade falls in the driven region. On the retreating side, more of the blade is in the stall region. A small section near the root experiences a reversed flow, therefore the size of the driven region on the retreating side is reduced.

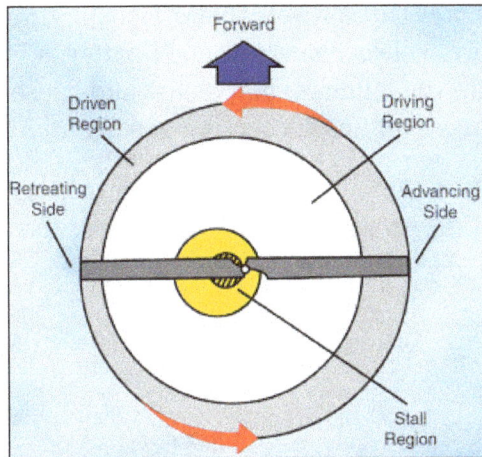

Blade regions in forward autorotation descent.

Launcher Aerodynamics

During the design phase of launchers, the aerodynamic characterization represents a fundamental contribution. Usually, it is accomplished by means a hybrid approach encompassing wind tunnel testing (WTT) and computational fluid dynamics (CFD) investigations. This combined design approach (i.e., WTT and CFD analyses) is extremely reliable in providing high quality data as input for launchers' sizing, performance evaluations, control, and staging dynamics. Indeed, launcher aerodynamics focuses on the assessment of the pressure and skin friction loads the atmosphere determines over the vehicle surface. As well known, these loads result in a global aerodynamic force that acts at the aeroshape center of pressure (CoP) which generally does not coincide with the vehicle center of gravity (CoG). As a result, the related aerodynamic moment acting at the CoG can lead to a stable or unstable behavior of the launcher to account for in the control software. Moreover, the analysis of the flowfield past the launcher is also fundamental to address the effects of aeroshape's structures and protrusions. Indeed, aeroshell steps and gaps determine local pressure (and convective heat flux) overshoots all along the ascent trajectory. This assessment is fundamental for launcher sizing and thermal protection design activities.

CFD Validation Study

In the last years, CFD has played an important role in hypersonics being able to address particular design issues, such as the well-known SSI and SWIBLI. These flowfield features occur whenever different shocks interact each other or with the boundary layer when a shock impinges on a wall, respectively. For launchers, SSI and SWIBLI phenomena typically take place in the flowfield region within launcher main body and boosters.

In this figure, the fairings bow shock meets that of booster, thus originating a SSI. This interaction results in more or less complex shock patterns including shear-layers or jets, which can impinge on the launcher aeroshape and cause local pressure and heat flux overshoots, well in excess of those occurring at stagnation points.

On the other hand, SWBLI occurs, for instance, when the shock resulted from the SSI meets the launcher wall, thus promoting boundary layer separation and transition.

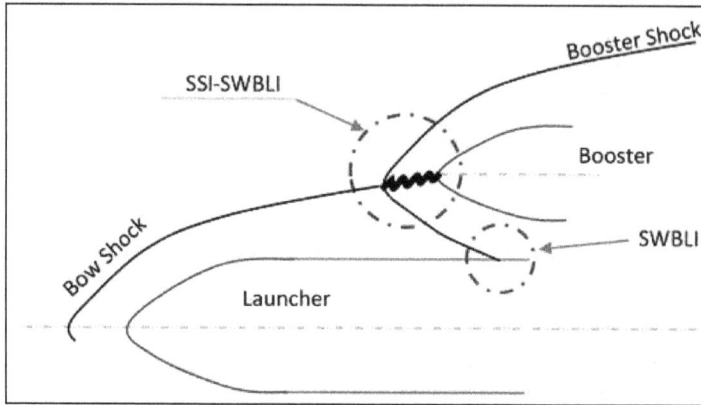

Example of SSI and SWBLI for launchers.

As a result, SSI and SWBLI demand accurate prediction for a reliable and affordable aerothermal design of launcher vehicles.

This configuration, in fact, is a benchmark as it presents unique flow patterns typical of SSI and SWIBLI. In particular, the experiment of Swantek and Austin was selected and numerically rebuilt. The test bed geometry is shown in figure. It is a double wedge with $\theta_1 = 30°$ and $\theta_2 = 55°$ where the lengths of the first and second face are $L_1 = 50.8$ mm and $L_2 = 25.4$ mm, respectively.

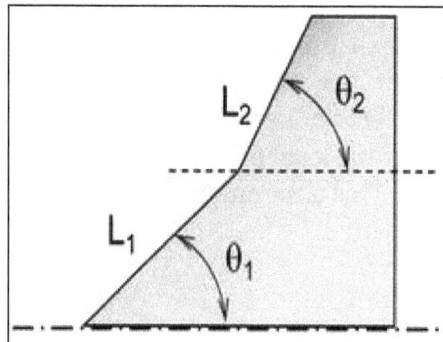

Test bed configuration with quotes.

Along with the center of the model 19 coaxial thermocouple gauges at 16 different streamwise locations are mounted. Therefore, several experimental data exist for numerical-to-experimental comparisons. The test campaign was performed by using high enthalpy air at the free-stream conditions summarized in table.

Table: Free-stream conditions of experiment.

Parameter	M7_8
Stagnation enthalpy (MJ/kg)	8.0
Mach	7.14
Static temperature (K)	710
Static pressure (kPa)	0.780
Velocity (m/s)	3812
Density (kg/m³)	0.0038
Unit Reynolds number (10^6/m)	0.435

The numerical rebuilding was carried out by means of a steady-state two-dimensional Reynolds-averaged Navier-Stokes (RANS) simulation performed with the commercial CFD tool Fluent. Air was modeled with a five species chemistry mixture (N_2, N, O_2, O, NO) in thermo-chemical non-equilibrium conditions. Turbulence has been taken into account with the k–ω SST model. The wall was assumed isothermal (T_w = 298 K) and noncatalytic; while, in order to take into account the effects of the boundary layer transition, a trade-off analysis was undertaken in order to determine a proper flow transition location (x_{tr}) to fix along with the first ramp. Results highlighted that x_{tr} = 58% L_1 is a viable option.

A structured multiblock mesh of 433 × 707 points was considered to solve for complex flow structure past the test bed. In particular, a great deal of care was taken in grid development. In fact, the distribution of grid points has been dictated by the level of resolution desired in various areas of the computational domain such as SSI, triple points, shear layer and recirculation region. An example of the computational grid is provided in figure.

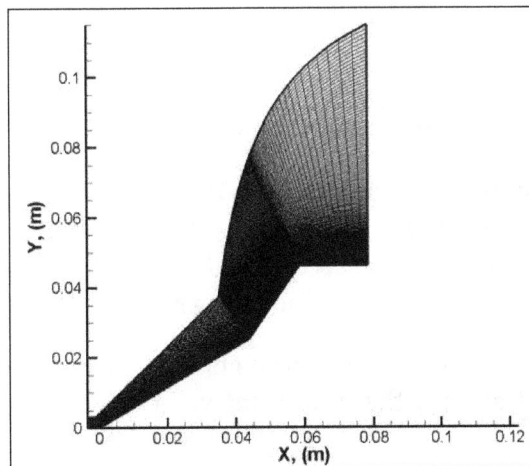

An example of the computational grid.

As far as numerical results are concerned, figure shows the qualitative comparison between experimental data (i.e., Schlieren image) and the Mach isolines.

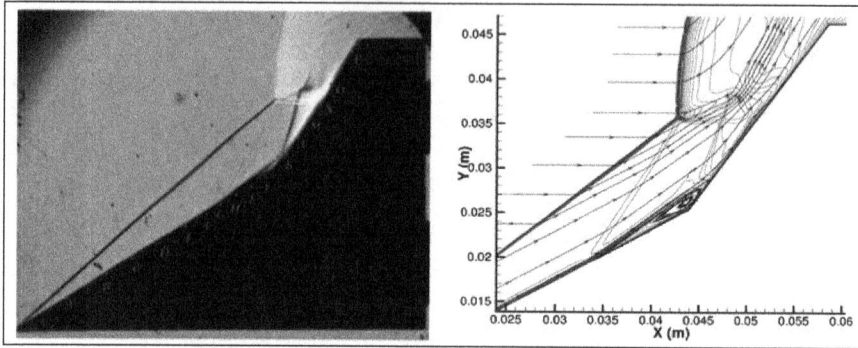

Comparison between Schlieren image and Mach isolines with streamtraces.

As one can see, CFD results compare rather well with the Schlieren. Indeed, the numerical flowfield presents the same structure as pointed out by the experimental data, as the triple point, due to a strong shock that originates ahead the recirculation bubble, the reattachment shock and the shear layer.

Results comparison in terms of pressure and heat flux distribution is presented in figure, where measures available for the heat flux are also provided.

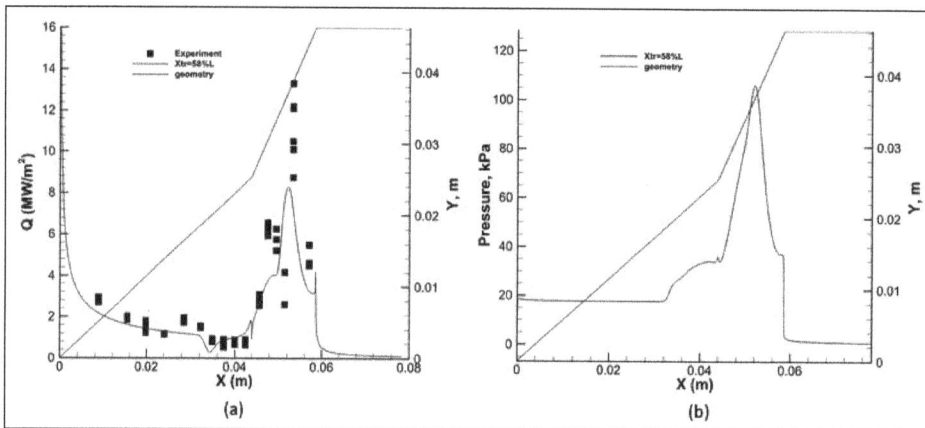

Heat flux and pressure profiles. Comparisons with experimental data.

As shown, the computed heat transfer is within the experimental uncertainty upstream of the separation point at x_{tr} = 58% L_1 (i.e., x = 27 mm); while rather good agreement with experiment is observed over the second wall of the double wedge, where the heat flux and pressure overshoots take place due to the shear layer impingement.

In particular, figure points out that CFD predicts well the recirculation bubble and the peak heat transfer location, but the numerical value is about 43% of that measured during the experiment.

Regarding pressure distribution, it is noticeable the pressure increase behind the separation shock on the first ramp. Then, a pressure overshoot, located just downstream of the reattachment point, is predicted on the second ramp. This is typical for the Edney type IV interaction. After the peak, the pressure suddenly drops toward the asymptotic pressure due to the strong expansion at the end of ramp.

Launcher Aerodynamic Appraisal

The launcher vehicle features a hummer head cylinder, as main body, with two boosters. Non-dimensional aeroshape sizes are also reported in figure, being L the launcher height. As shown, the aeroshape under investigation also features a central core stage with a remarkable boat-tail configuration, which ends in correspondence of booster stage. The fairing diameter is 16% launcher height, while that of booster is equal to 0.076 L. The booster length is 40% of whole launcher's height.

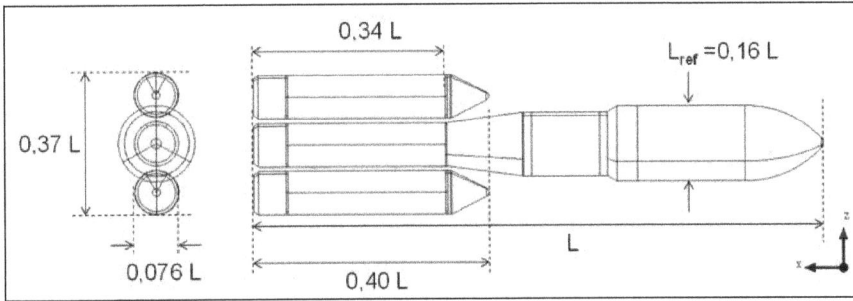

The launcher configuration.

Aerodynamic data for launcher are provided in the Body Reference Frame (BRF), as illustrated in figure. In this figure, aerodynamic force and moment coefficients are also provided, with sign convention according to the ISO norm. 1151.

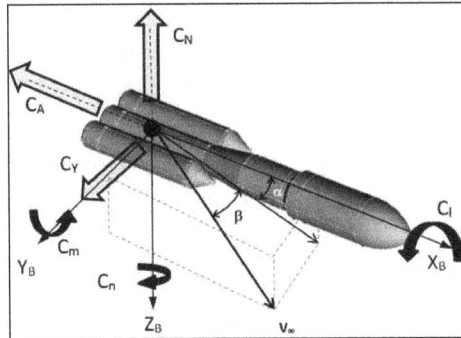

The body reference frame.

The global aerodynamic force $\vec{F}$ and moment $\vec{M}$ acting on the launcher are expressed in BRF as follows:

$$\vec{F} = \left(-F_A \hat{i} + F_Y \hat{j} - F_N \hat{k} \right) = S_{ref} q_\infty \left(-C_A \hat{i} + C_Y \hat{j} + C_N \hat{k} \right)$$

$$\vec{M} = \left(M_l \hat{i} + M_m \hat{j} - M_n \hat{k} \right) = S_{ref} q_\infty L_{ref} \left(C_l \hat{i} + C_m \hat{j} + C_n \hat{k} \right)$$

Where CA is the axial force coefficient, CY the transverse force coefficient, CN the normal force coefficient, Cl = CMx the rolling moment coefficient, Cm = CMy the pitching moment coefficient, Cn = CMz the yawing moment coefficient, $\left(\hat{i},\hat{j},\hat{k}\right)$ are the reference unit vectors, Sref the reference surface, Lref the reference length, and $q\infty$ the free-stream dynamic pressure.

The definition of force and moment coefficients is:

$$C_i = \frac{F_i}{S_{ref}q_\infty} \quad i = A, Y, N$$

$$C_i = \frac{M_i}{S_{ref}q_\infty L_{ref}} \quad i = l, m, n$$

where $\rho\infty$ is the atmospheric density and V∞ the speed relative to air, and the reference quantities are:

$$L_{ref} = 0.16\,L$$

$$S_{ref} = \frac{\pi L_{ref}^2}{4}$$

The present preliminary assessment, however, focuses on the longitudinal aerodynamic only, i.e., CA, CN and Cm are addressed. Aerodynamic coefficients are important at system level for the assessment of launcher general loading determinations, performances and, as well as, control. For instance, performances studies use the axial force coefficient CA since this aerodynamic force opposes to the vehicle movement. Further, the launcher control needs the evaluation of the aerodynamic moment at the CoG since the control software changes the rocket's thrust direction in order to null global incidence of the vehicle, except during maneuvers. Anyway, considering that propellants are constantly consumed along the ascent flight, the CoG location is continuously changing too. Therefore, it is preferred to provide aerodynamic moments at a conventional location, namely moment reference center (MRC). The relationship for the pitching moment coefficient evaluation, passing from MRC to CoG, reads:

$$\left(C_m\right)_{CoG} = \left(C_m\right)_{MRC} + C_N \frac{\Delta x}{L_{ref}} - C_A \frac{\Delta z}{L_{ref}}$$

where, $\Delta x = xCoG - xMRC$ and $\Delta z = zCoG - zMRC$ are evaluated in the Layout Reference Frame (LRF), as shown in figure.

The flow regime investigated for launcher aerodynamic appraisal during ascent encompasses subsonic, transonic-supersonic and hypersonic regimes.

The range $0.5 \leq M_\infty \leq 5$ is investigated. Indeed, launcher aerodynamics has been addressed considering four Mach numbers, namely 0.5, 1.1, 2.5, and 5, at three angle of attacks, i.e., α = 0, 5, and 7°, as summarized by the CFD test matrix in table. Therefore, Eulerian and

Navier-Stokes 3D CFD computations have been carried out on several unstructured hybrid meshes and in motor-off (i.e., without the effect of rocket plume) conditions.

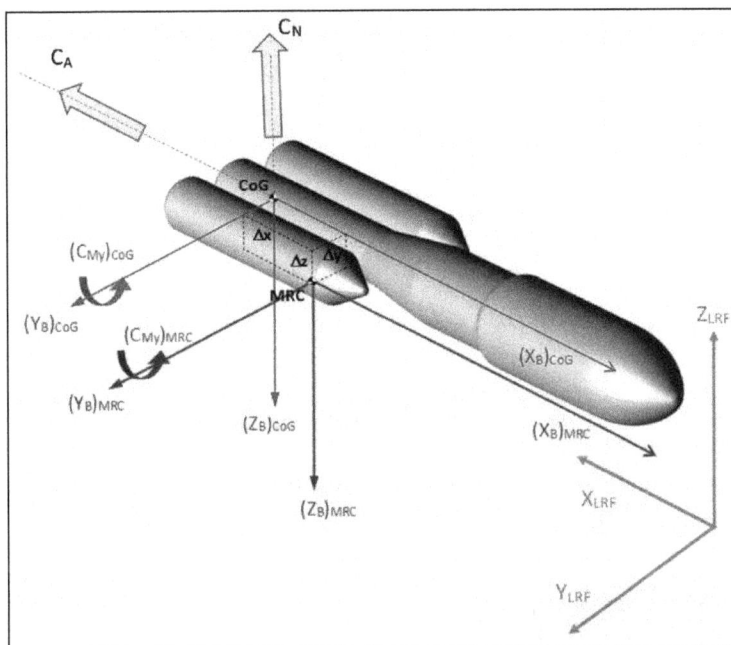

The layout reference frame.

Table: The CFD test matrix.

α (°)	Mach			
	0.5	1.1	2.5	5
0	E	E	E	E
5	E	E	E, NS	E, NS
7	E	E	E	E

where, E: Eulerian CFD; NS: Navier-Stokes CFD.

Engineering-based aerodynamic analyses were also performed by using a 3D Panel Method (PM) code, namely Surface Impact Method (SIM), developed by CIRA. This tool is able to accomplish the supersonic and hypersonic aerodynamic and aerothermodynamic analyses of complex vehicles configuration by using simplified approaches as local surface inclination methods and approximate boundary-layer methods, respectively. Surface impact methods typical of Hypersonics, are Newtonian, Modified Newtonian, Tangent cone and Tangent Wedge theories.

A typical mesh surface that has been used for the engineering-level computations is shown in figure. Some engineering-based aerodynamic results for axial and normal force coefficient are provided in both figures, respectively.

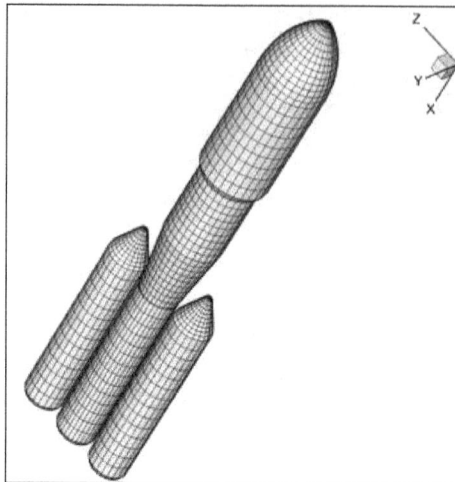

Panel code mesh for sup-hypersonic aerodynamics.

On the other hand, the mesh domains for subsonic and sup-hypersonic speed flow simulations are shown in figures, respectively.

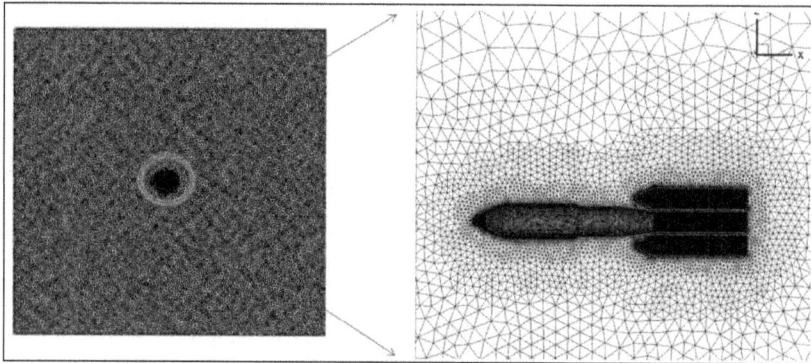

Overview of the hybrid mesh domain for subsonic speed.

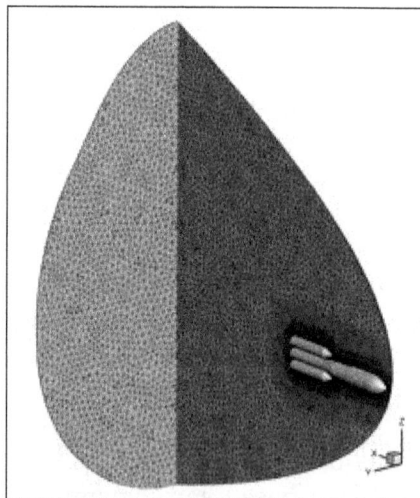

Overview of a hybrid mesh domain for sup-hypersonic CFD simulations.

As on can see, a square brick wide 20 body length upstream, downstream, upward and downward the launcher is considered to assure farfield unperturbed free-stream flow conditions at subsonic speed. Indeed, in this flow regime (i.e., elliptic flow), disturbances due to the body influence flow everywhere since they are propagated upstream via molecular collisions at approximately the speed of sound. Therefore, the computational domain must be wide enough to avoid interferences between flowfield and farfield boundary conditions.

At supersonic speed, however, a shock wave appears at launcher leading edge (i.e., hyperbolic flowfield) because of, when flow moves faster than the speed of sound, disturbances cannot work their way upstream but coalesce forming a standing wave, namely bow shock. As a result, the computational domain is quite narrow, as shown in figure.

CFD results of the preliminary assessment of launcher aerodynamics are summarized from figures. For instance, figure shows the pressure distribution expected on the surface of the launcher flying at $M_\infty = 0.5$ and $\alpha = 5°$. Flow compression that takes place for this flight conditions at the stagnation regions of launcher fairings and of boosters' conical forebody is clearly shown. A recompression zone at the beginning of the cylindrical trunk, just after the fairings, and on that close to the boosters' forebody can be noted as well.

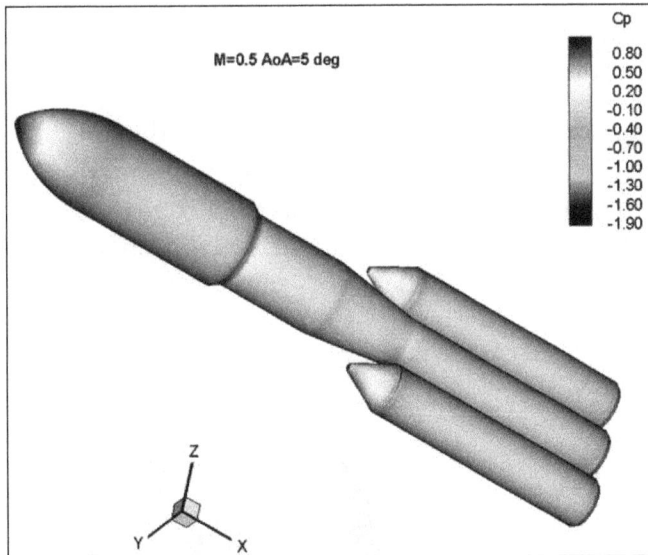

Pressure coefficient at $M_\infty = 0.5$ and $\alpha = 5°$.

Results for numerical investigations at higher Mach numbers are provided in figure. Here, an overview of pressure coefficient (Cp) distribution on launcher symmetry plane and surface is provided for $M_\infty = 2.5$ and $\alpha = 5°$.

Flow streamtraces on the symmetry plane are reported as well. This CFD computation is carried out with SST k-ω turbulence flow model and for cold wall boundary condition (i.e., Tw = 300 K).

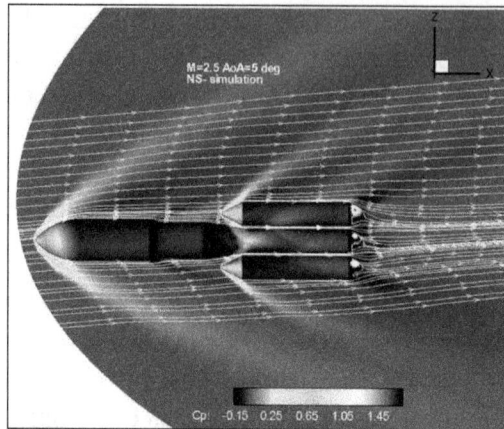

Overview of Cp distribution on symmetry plane and launcher at $M_\infty = 2.5$ and $\alpha = 5°$.

Results in figure highlight a complex flowfield past the launcher due to the flow separation bubble at fairing boat-tail and the effect of fuselage/booster SSI and SWIBLI. For instance, after compression at conical flare of main fairings the flow undergoes to expansions that align it along with the constant cross section part of hammerhead. Hence, at the end of fairings another strong expansion takes place to accommodate the flow to the variation in launcher cross section (i.e., narrow cross section due to fairing boat-tail). Then, a shock wave arises at the beginning of the cylindrical trunk, just after the fairings, to redirect the flow along with the launcher wall.

Flow complexity increases further in the region close to the boosters leading edges, as also shown in figure. This figure provides an overview of pressure coefficient distribution on launcher aeroshape with skin friction lines. As one can see, in the region close to the boosters' leading edges, complex SSI and SWIBLI phenomena take place. They result in higher thermo-mechanical loads (i.e., local pressure and thermal overshoots) on the launcher wall that must be carefully addressed in the vehicle design.

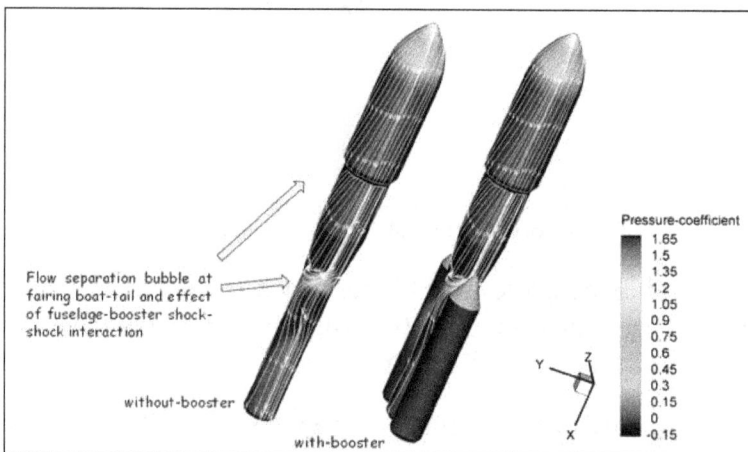

Contours of Cp with skin friction lines on launcher aeroshape at $M_\infty = 2.5$ and $\alpha = 5°$.

The effect of SSI between launcher and booster at $M_\infty = 5$ and $\alpha = 0°$ flight conditions is clearly highlighted by the pressure overshoots shown at about x = 26 m in figure. As one can see, also at those flight conditions complex flowfield interaction phenomena are expected.

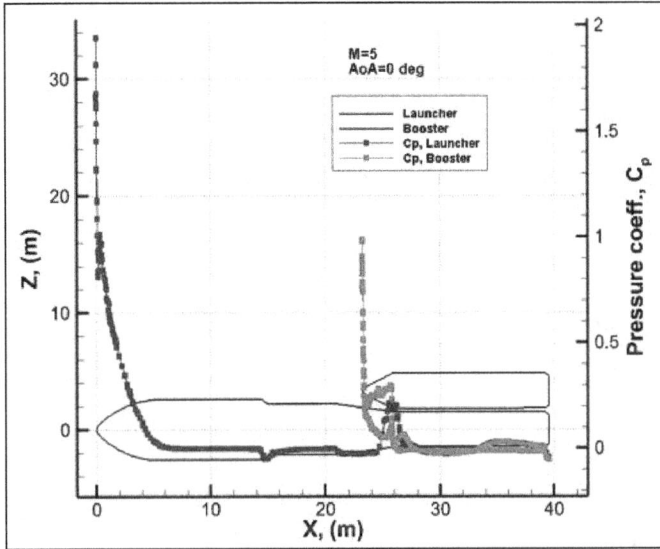

Profiles of Cp on launcher and booster centerlines at M_∞ =5 and α = 0°.

As far as aerodynamic coefficients are concerned, results for launcher axial force, normal force and pitching moment coefficients are summarized from figures. For instance, figure shows the axial force coefficient versus Mach number at different AoA, namely α = 0, 5, and 7°. As one can see, C_A does not significantly change passing from 0 to 7° AoA at each considered Mach number.

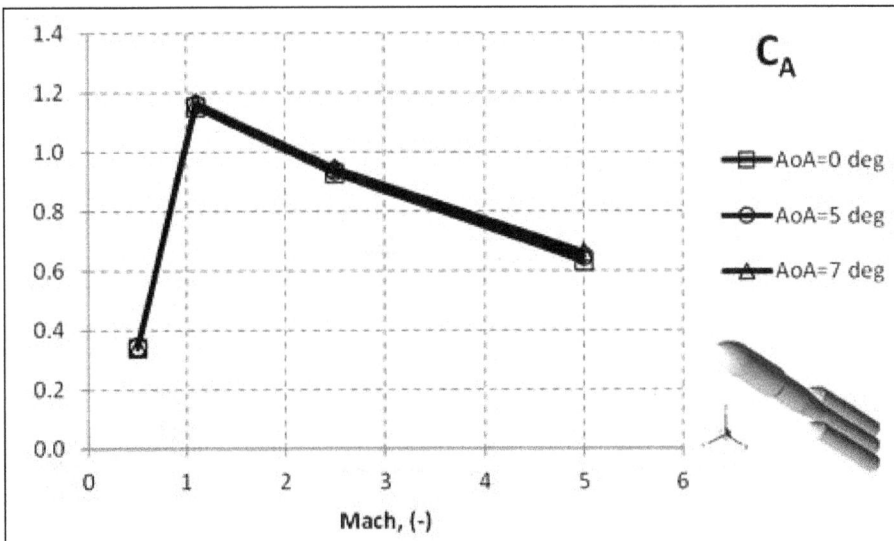

Axial force coefficient versus Mach at different AoA, namely α = 0, 5, and 7°.

On the contrary, the effect of flow compressibility is remarkable, as expected. Indeed, the strong increase to which undergoes the axial aerodynamic force, when M_∞ becomes transonic, is due to the wave drag contribution, as expected. Nevertheless, this contribution tends to be less strong as Mach number goes toward hypersonic speed conditions considering that the shock becomes weak due to the streamlined vehicle aeroshape (i.e., high inclined shock to assure a narrow shock layer).

The variation of C_A versus the angle of attack, α, at $M_\infty = 2.5$ and 5 is provided in figure, where a comparison between SIM and CFD results is also available. As shown, engineering and numerical results compare rather well, thus confirming the reliability of the panel methods outcomes.

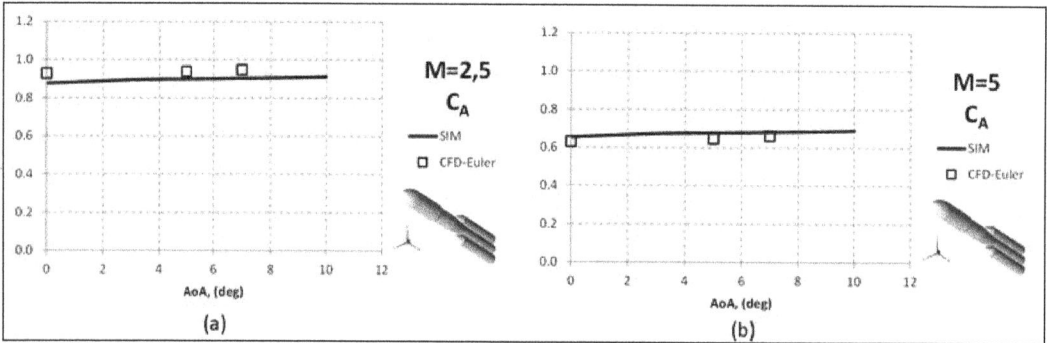

CA versus AoA at $M_\infty = 2.5$ and 5. Comparison between PM and CFD results.

Regarding normal force coefficient results, figure highlights that, for each Mach number, C_N features a quite linear slope as α increases up to 7° AoA. In addition, in this case, compressibility effect influences launcher normal force by means of different curve slopes for each Mach number. Results comparison between SIM and CFD is provided in figure at $M_\infty = 2.5$ and 5.

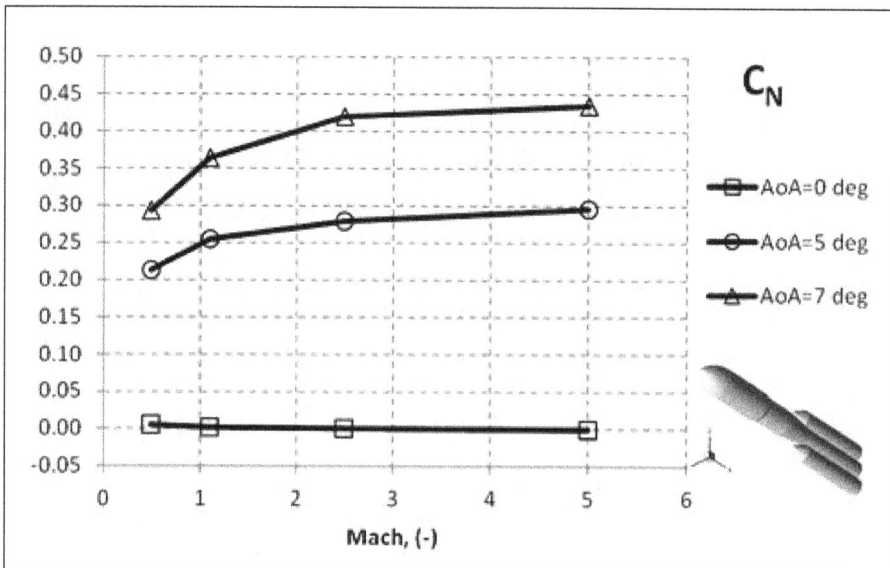

Normal force coefficient versus Mach at different AoA, namely $\alpha = 0$, 5, and 7°.

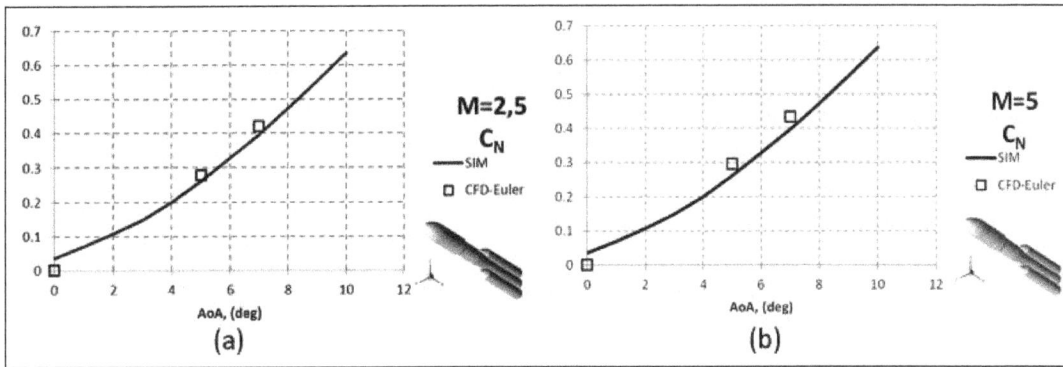

CN versus AoA at $M_\infty = 2.5$ and 5. Comparison between PM and CFD results.

As one can see, the reliability of the panel methods outcomes is still confirmed.

The vehicle pitching moment coefficient features a behavior quite close to that described for the C_N, but with a strong pitch down detected moving toward $M_\infty = 5$.

Note that both C_N and C_m at $\alpha = 0°$ are null due to the symmetric launcher aeroshape.

Finally, the axial coefficient breakdown at $M_\infty = 5$ and $\alpha = 0°$ is shown in figure. Here the lumped contributions of launcher fairings, boat-tail, core, cylinder, and base, as well as of booster fuse and base are recognized. As one can see, launcher fairings contribute to about 68% of total drag coefficient; while this percentage for booster fuselage and base is close to 21 and 5%, respectively.

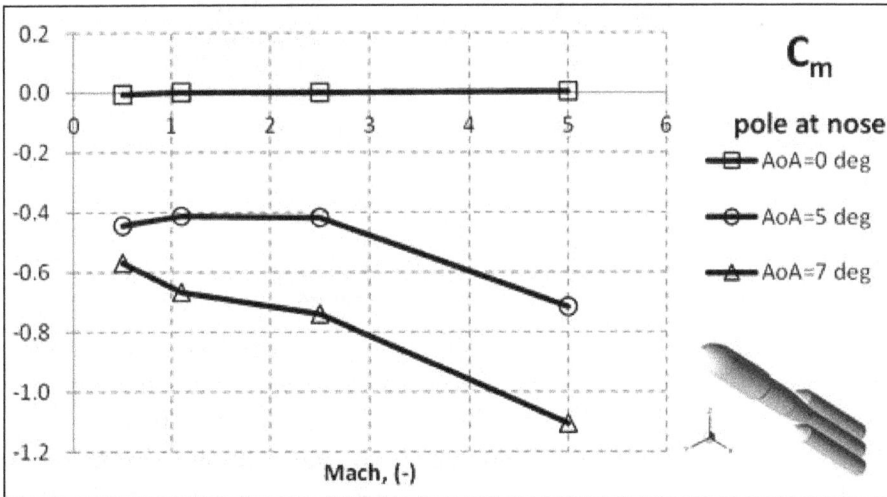

Pitching moment coefficient versus Mach at different AoA, namely $\alpha = 0$, 5, and 7°.

The goal is to address the preliminary aerodynamic database of a typical launch vehicle configuration as input for performances evaluations as well as launcher control, sizing, and staging dynamics. To this end, both reliable engineering-based and steady-state computational fluid dynamics, with both Euler and Navier-Stokes approximations, are carried out at several Mach numbers, vehicle attitude conditions, and in motor-off conditions. In particular, launcher aerodynamic performance is provided in terms of axial, normal and pitching moment coefficients. Numerical results point out that the axial force coefficient does not significantly change passing from 0 to 7° angle of attack at each considered Mach number; while the effect of flow compressibility is remarkable. Regarding normal force coefficient, results highlight that, for each Mach number, it features a quite linear slope as the angle of attack increases up to 7°. Finally, the behavior of the vehicle pitching moment coefficient is quite close to that described for the normal force coefficient, but a strong pitch down is detected when launcher speed becomes hypersonic.

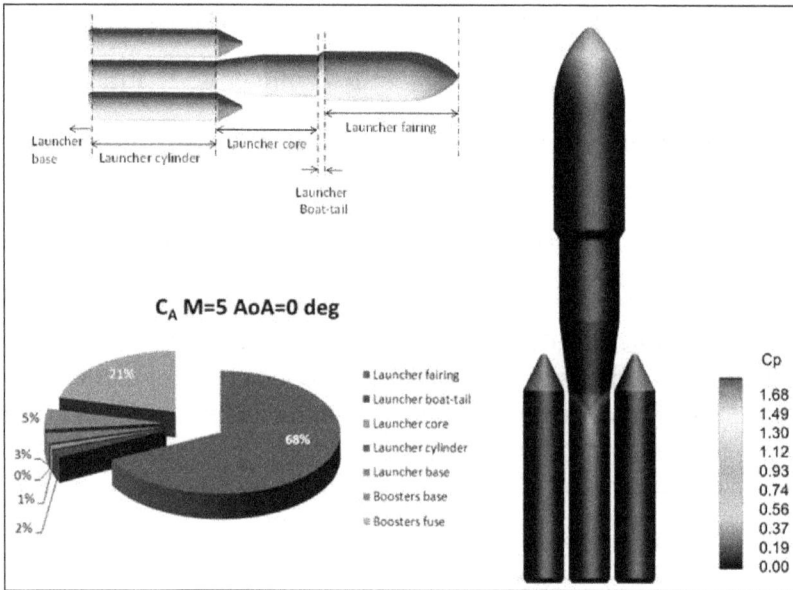

CA breakdown at $M\infty = 5$ and $\alpha = 0°$.

Wind Turbines Aerodynamics

A wind turbine is a device that extracts kinetic energy from the wind and converts it into mechanical energy. Therefore wind turbine power production depends on the interaction between the rotor and the wind. So the major aspects of wind turbine performance like power output and loads are determined by the aerodynamic forces generated by the wind. These can only be understood with a deep comprehension of the aerodynamics of steady state operation.

Aerodynamics of HAWTs

The analysis of the aerodynamic behavior of wind turbines can be started without any specific turbine design just by considering the energy extraction process. A simple model, known as actuator disc model, can be used to calculate the power output of an ideal turbine rotor and the wind thrust on the rotor. Additionally more advanced methods including momentum theory, blade element theory and finally blade element momentum (BEM) theory are introduced. BEM theory is used to determine the optimum blade shape and also to predict the performance parameters of the rotor for ideal, steady operating conditions. Blade element momentum theory combines two methods to analyze the aerodynamic performance of a wind turbine. These are momentum theory and blade-element theory which are used to outline the governing equations for the aerodynamic design and power prediction of a HAWT rotor. Momentum theory analyses the momentum balance on a rotating annular stream tube passing through a turbine and blade-element theory examines the forces generated by the aerofoil lift and drag coefficients at various sections along the blade. Combining these theories gives a series of equations that can be solved iteratively.

Actuator Disc Model

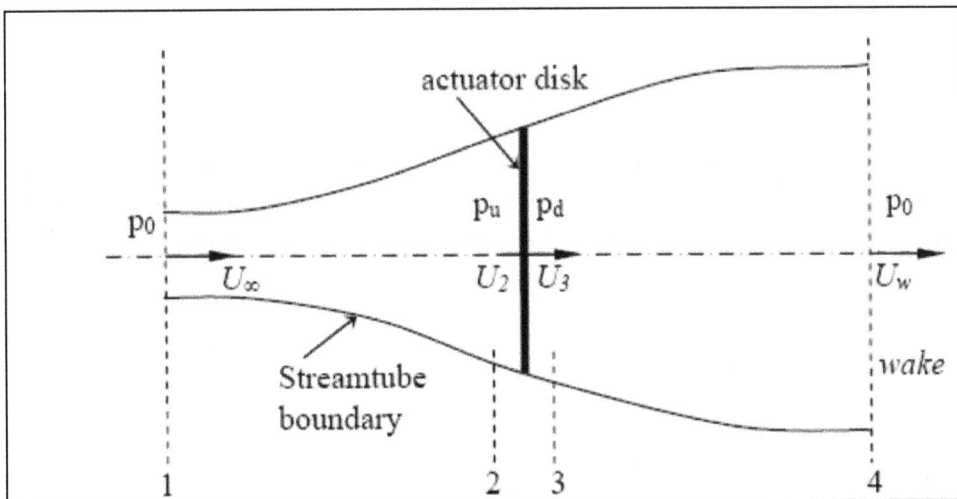

Actuator Disk Model.

The analysis of the aerodynamic behavior of wind turbines can be started without any specific turbine design just by considering the energy extraction process. The simplest model of a wind turbine is the so-called actuator disc model where the turbine is replaced by a circular disc through which the airstream flows with a velocity U_∞ and across which there is a pressure drop from p_u to p_d At the outset, it is important to stress that the actuator disc theory is useful in discussing overall efficiencies of turbines but it cannot be utilized to design the turbine blades to achieve a desired performance. Actuator disc model is based on the assumptions like no frictional drag, homogenous, incompressible, steady state fluid flow, constant pressure increment or thrust per unit area over the disk, continuity of velocity through the disk and an infinite number of blades.

The analysis of the actuator disk theory assumes a control volume in which the boundaries are the surface walls of a stream tube and two cross-sections. In order to analyze this control volume, four stations (1: free-stream region, 2: just before the blades, 3: just after the blades, 4: far wake region) need to be considered. The mass flow rate remains the same throughout the flow. So the continuity equation along the stream tube can be written as:

$$\rho A_\infty U_\infty = \rho A_d U_d = \rho A_w U_w$$

Assuming the continuity of velocity through the disk gives:

$$U_2 = U_3 = U_R$$

For steady state flow the mass flow rate can be obtained using:

$$\dot{m} = \rho A U_R$$

Applying the conservation of linear momentum equation on both sides of the actuator disk gives:

$$T = \dot{m}\left(U_\infty - U_w\right)$$

Since the flow is frictionless and there is no work or energy transfer is done, Bernoulli equation can be applied on both sides of the rotor. If we apply energy conservation using Bernoulli equation between station above four equation can be obtained respectively.

$$p_d + \frac{1}{2}\rho U_R^2 = p_o + \frac{1}{2}\rho U_w^2$$

$$p_o + \frac{1}{2}\rho U_\infty^2 = p_u + \frac{1}{2}\rho U_R^2$$

Combining Equations: $p_d + \frac{1}{2}\rho U_R^2 = p_o + \frac{1}{2}\rho U_w^2$ and $p_o + \frac{1}{2}\rho U_\infty^2 = p_u + \frac{1}{2}\rho U_R^2$ gives the pressure decrease p' as:

$$p' = \frac{1}{2}\rho\left(U_\infty^2 - U_w^2\right)$$

Also the thrust on the actuator disk rotor can be expressed as the sum of the forces on each side,

$$T = Ap'$$

where,

$$p' = \left(p_u - p_d\right)$$

Substituting equation $p' = \frac{1}{2}\rho\left(U_\infty^2 - U_w^2\right)$ into equation $T = Ap'$ gives the thrust on the disk in more explicit form:

$$T = \frac{1}{2}A\rho\left(U_\infty^2 - U_w^2\right)$$

Combining equations $\dot{m} = \rho A U_R$, $T = \dot{m}\left(U_\infty - U_w\right)$ and $T = \frac{1}{2}A\rho\left(U_\infty^2 - U_w^2\right)$ the velocity through the disk can be obtained as:

$$U_R = \frac{U_\infty + U_w}{2}$$

Defining the axial induction factor α as in equation:

$$a = \frac{U_\infty - U_R}{U_\infty}$$

gives Equation:

$$U_R = U_\infty\left(1 - a\right)$$

$$U_w = U_\infty\left(1 - 2a\right)$$

To find the power output of the rotor equation can be used.

$$P = TU_R$$

By substituting equation $T = \frac{1}{2}A\rho\left(U_\infty^2 - U_w^2\right)$ into $P = TU_R$ gives the power output based on the momentum balance on both sides of the actuator disk rotor in more explicit form.

$$P = \frac{1}{2}\rho\left(U_\infty^2 - U_W^2\right)U_R$$

Also substituting equations $U_R = U_\infty\left(1 - a\right)$ and $U_w = U_\infty\left(1 - 2a\right)$ into equation $P = TU_R$ gives:

$$P = 2\rho A a U_\infty^3\left(1 - a\right)^2$$

Finally the performance parameters of a HAWT rotor (power coefficient C_p, thrust coefficient C_T, and the tip-speed ratio λ) can be expressed in dimensionless form which is given in Equation below:

$$C_P = \frac{2P}{\rho U_\infty^3 \pi R^2}$$

$$C_T = \frac{2T}{\rho U_\infty^2 \pi R^2}$$

$$\lambda = \frac{R\Omega}{U_\infty}$$

Substituting $P = 2\rho A a U_\infty^3 (1-a)^2$ into $C_P = \frac{2P}{\rho U_\infty^3 \pi R^2}$, the power coefficient of the rotor can be rewritten as:

$$C_p = 4a(1-a)^2$$

Also using the equations $T = \frac{1}{2} A\rho (U_\infty^2 - U_w^2)$, $U_R = U_\infty (1-a)$ and $P = 2\rho A a U_\infty^3 (1-a)^2$ the axial thrust on the disk can be rewritten as:

$$T = 2Aa\rho(1-a)U_\infty^2$$

Finally substituting equation $T = 2Aa\rho(1-a)U_\infty^2$ into equation,

$$C_T = \frac{2T}{\rho U_\infty^2 \pi R^2}$$

gives the thrust coefficient of the rotor as:

$$C_T = 4a(1-a)$$

Rotating Annular Stream Tube Analysis

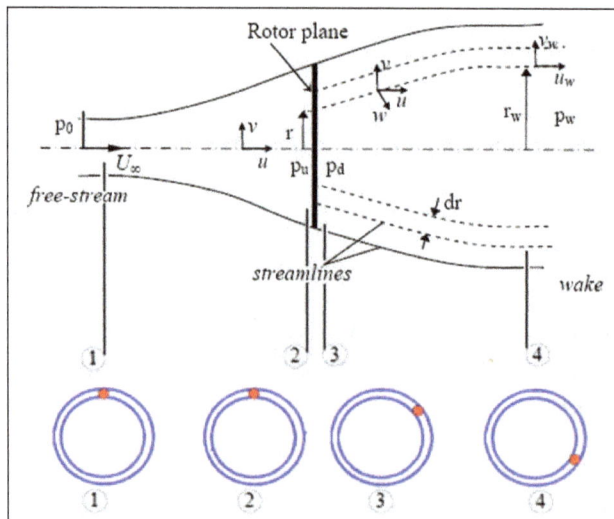

Title Rotating Annular Stream Tube Analysis.

Thus far the method is developed on the assumption that there was no rotational motion. To extend the method developed, the effects of this rotational motion needs to be included so it is necessary to modify the qualities of the actuator disk by assuming that it can also impart a rotational component to the fluid velocity while the axial and radial components remain unchanged. Using a rotating annular stream tube analysis, equations can be written that express the relation between the wake velocities (both axial and rotational) and the corresponding wind velocities at the rotor disk.

This analysis considers the conservation of angular momentum in the annular stream tube. If the condition of continuity of flow is applied for the annular element taken on the rotor plane can be written:

$$u_w r_w dr_w = urdr$$

Applying the conservation of the angular momentum on upstream and the wake region of the flow domain gives:

$$w_w r_w^2 = wr^2$$

Also the torque caused by the angular momentum balance on the differential annular element can be obtained using:

$$dQ = \rho uwr^2 dA$$

where $dA = 2r\pi dr$. Also applying the Bernoulli equation between station between above three equation and 4 gives Bernoulli's constants as:

$$H_o = p_o + \frac{1}{2}\rho U_\infty^2 = p_u + \frac{1}{2}\rho\left(u^2 + v^2\right)$$

$$H_1 = p_d + \frac{1}{2}\rho\left(u^2 + v^2 + w^2 r^2\right) = p_w + \frac{1}{2}\rho\left(u_w^2 + r_w^2 w_w^2\right)$$

And taking the difference between these constants gives:

$$H_o - H_1 = p' - \frac{1}{2}\rho\left(w^2 r^2\right)$$

Which means the kinetic energy of the rotational motion given to the fluid by the torque of the blade is equal to $-(1/2)\rho\left(w^2 r^2\right)$. So the total pressure head between both sides of the rotor becomes:

$$p_o - p_w = \frac{1}{2}\rho\left(U_w^2 - U_\infty^2\right) + \frac{1}{2}w_w^2\rho r_w^2 + \left(H_o - H_1\right)$$

$$= \frac{1}{2}\rho\left(u_w^2 - U_\infty^2\right) + \frac{1}{2}\rho\left(w_w^2 r_w^2 - w^2 r^2\right) + p'$$

Applying the Bernoulli's equation between station just above equation gives the pressure drop as:

$$p' = \frac{1}{2}\rho\left[-\Omega^2 + (\Omega + w)^2\right]r^2 = \rho\left(\Omega + \frac{w}{2}\right)wr^2$$

Substituting this result into the equation:

$$p_o - p_w = \frac{1}{2}\rho\left(U_w^2 - U_\infty^2\right) + \frac{1}{2}w_w^2\rho r_w^2 + (H_o - H_1)$$

$$= \frac{1}{2}\rho\left(u_w^2 - U_\infty^2\right) + \frac{1}{2}\rho\left(w_w^2 r_w^2 - w^2 r^2\right) + p'$$

gives:

$$p_o - p_w = \frac{1}{2}\rho\left(u_w^2 - U_\infty^2\right) + \rho\left(\Omega + \frac{w}{2}\right)r_w^2 w_w$$

In above station, the pressure gradient can be written as:

$$\frac{dp_w}{dr_w} = \rho r_w w_w^2$$

Differentiating equation $p_o - p_w = \frac{1}{2}\rho\left(u_w^2 - U_\infty^2\right) + \rho\left(\Omega + \frac{w}{2}\right)r_w^2 w_w$ relative to r_w and equating to equation just above gives:

$$\frac{1}{2}\frac{d}{dr_w}\left(U_\infty^2 - u_w^2\right) = (\Omega + w_w)\frac{d}{dr_w}\left(r_w^2 w_w\right)$$

The equation of axial momentum for the given annular blade element in differential form can be written as:

$$dT = \rho u_w(U_\infty - u_w)dA_w + (p_o - p_w)dA_w$$

Since, $dT = p'dA$, the previous equation can be written as:

$$dT = \rho\left(\Omega + \frac{w}{2}\right)wr^2 dA$$

Finally, combining equation:

$$u_w r_w dr_w = u r dr$$

$$p_o - p_w = \frac{1}{2}\rho\left(U_w^2 - U_\infty^2\right) + \frac{1}{2}w_w^2\rho r_w^2 + \left(H_o - H_1\right)$$

$$= \frac{1}{2}\rho\left(u_w^2 - U_\infty^2\right) + \frac{1}{2}\rho\left(w_w^2 r_w^2 - w^2 r^2\right) + p'$$

$$dT = \rho u_w\left(U_\infty - u_w\right)dA_w + \left(p_o - p_w\right)dA_w$$

and

$$dT = \rho\left(\Omega + \frac{w}{2}\right)wr^2 dA$$

gives:

$$\frac{1}{2}\left(U_\infty - u_w\right)^2 = \left(\frac{\Omega + w_w/2}{u_w} - \frac{\Omega + w/2}{U_\infty}\right)u_w r_w^2 w_w$$

An exact solution of the stream-tube equations can be obtained when the flow in the slipstream is not rotational except along the axis which implies that the rotational momentum wr² has the same value for all radial elements. Defining the axial velocities as $u = U_\infty\left(1-a\right)$ and $u_w = U_\infty\left(1-b\right)$.

$$a = \frac{b}{c}\left[1 - \frac{(1-a)b^2}{4\lambda^2(b-a)}\right]$$

Also the thrust on the differential element is equal to:

$$dT = 2\rho u\left(u - U_\infty\right)dA$$
$$= 4\pi\rho U_\infty^2 a\left(1-a\right)r dr$$

Using equation:

$$p' = \frac{1}{2}\rho\left[-\Omega^2 + \left(\Omega + w\right)^2\right]r^2 = \rho\left(\Omega + \frac{w}{2}\right)wr^2$$

Equation $dT = 2\rho u\left(u - U_\infty\right)dA = 4\pi\rho U_\infty^2 a\left(1-a\right)r dr$ can be rewritten as:

$$dT = p'dA$$
$$= 2\pi\rho\left(\Omega + w/2\right)wr^3 dr$$

If the angular induction factor is defined as $a' = \dfrac{w}{2\Omega}$, then dT becomes:

$$dT = 4\pi\rho\Omega^2 a'\left(1+a'\right)r^3 dr$$

In order to obtain a relationship between axial induction factor and angular induction factor, Equation $dT = p'dA = 2\pi\rho(\Omega + w/2)wr^3 dr$ and just above can be equated which gives:

$$\frac{a(1-a)}{a'(1+a')} = \frac{\Omega^2 r^2}{U_\infty^2} = \lambda_r^2$$

Using equation $dQ = \rho u w r^2 dA$ the torque on the differential element can be calculated as:

$$dQ = 4\pi\rho U_\infty \Omega a'(1-a)r^3 dr$$

The power generated at each radial element is given by $dP = \Omega dQ$. Substituting Equation $dQ = 4\pi\rho U_\infty \Omega a'(1-a)r^3 dr$ into equation gives:

$$dP = \frac{1}{2}\rho A U_\infty^3 \left[\frac{8}{\lambda^2} a'(1-a)\lambda_r^3 d\lambda_r \right]$$

Also the power coefficient for each differential annular ring can be written as:

$$dC_p = \frac{dP}{1/2\rho U_\infty^3 A}$$

Substituting,

$$dP = \frac{1}{2}\rho A U_\infty^3 \left[\frac{8}{\lambda^2} a'(1-a)\lambda_r^3 d\lambda_r \right]$$

into the Equation $dC_p = \dfrac{dP}{1/2\rho U_\infty^3 A}$ and integrating from hub tip speed ratio to the tip speed ratio gives power coefficient for the whole rotor.

$$C_p = \frac{8}{\lambda^2} \int_{\lambda_h}^{\lambda} a'(1-a)\lambda_r^3 d\lambda_r$$

By solving equation $\dfrac{a(1-a)}{a'(1+a')} = \dfrac{\Omega^2 r^2}{U_\infty^2} = \lambda_r^2$ for a' in terms of a equation can be obtained:

$$a' = -\frac{1}{2} + \frac{1}{2}\sqrt{1 + \frac{4}{\lambda_r^2} a(1-a)}$$

Solving the equations $C_p = \dfrac{8}{\lambda^2} \displaystyle\int_{\lambda_h}^{\lambda} a'(1-a)\lambda_r^3 \, d\lambda_r$ and $a' = -\dfrac{1}{2} + \dfrac{1}{2}\sqrt{1 + \dfrac{4}{\lambda_r^2} a(1-a)}$ together for the maximum possible power production gives:

$$\lambda_r^2 = \frac{(1-a)(4a-1)^2}{(1-3a)}$$

Also substituting equation $\lambda_r^2 = \dfrac{(1-a)(4a-1)^2}{(1-3a)}$ into Equation $\dfrac{a(1-a)}{a'(1+a')} = \dfrac{\Omega^2 r^2}{U_\infty^2} = \lambda_r^2$ gives the angular induction factor for maximum power in each annular ring.

$$a' = (1-3a)/(4a-1)$$

Differentiating Equation $\lambda_r^2 = \dfrac{(1-a)(4a-1)^2}{(1-3a)}$ with respect to axial induction factor at rotor plane, a relationship between r, dλ and da can be obtained.

$$2\lambda_r d\lambda_r = \left[6(4a-1)(1-2a)^2 / (1/3a)^2 \right] da$$

Finally, substituting the Equation,

$$\lambda_r^2 = \frac{(1-a)(4a-1)^2}{(1-3a)}, \ a' = (1-3a)/(4a-1),$$

$$2\lambda_r d\lambda_r = \left[6(4a-1)(1-2a)^2 / (1/3a)^2 \right] da$$

into the Equation $C_p = \dfrac{8}{\lambda^2} \displaystyle\int_{\lambda_h}^{\lambda} a'(1-a)\lambda_r^3 \, d\lambda_r$ gives the maximum power coefficient of the rotor:

$$C_{p,max} = \frac{24}{\lambda^2} \int_{a_1}^{a_2} \left[\frac{(1-a)(1-2a)(1-4a)}{(1-3a)} \right]^2 da$$

where, a_1 is the corresponding axial induction factor for $\lambda_r = \lambda_h$ and a_2 is the corresponding axial induction factor for $\lambda_r = \lambda$.

Blade Element Theory

Until this point the momentum theory is tried to be explained on account of HAWT rotor design but it does not consider the effects of rotor geometry characteristics like chord and twist distributions of the blade airfoil. For this reason blade element theory needs to be added to the design method. In order to apply blade element analysis, it is assumed that the blade is divided into N sections. This analysis is based on some assumptions including no aerodynamic interactions between different blade elements and the forces on the blade elements are solely determined by the lift and drag coefficients.

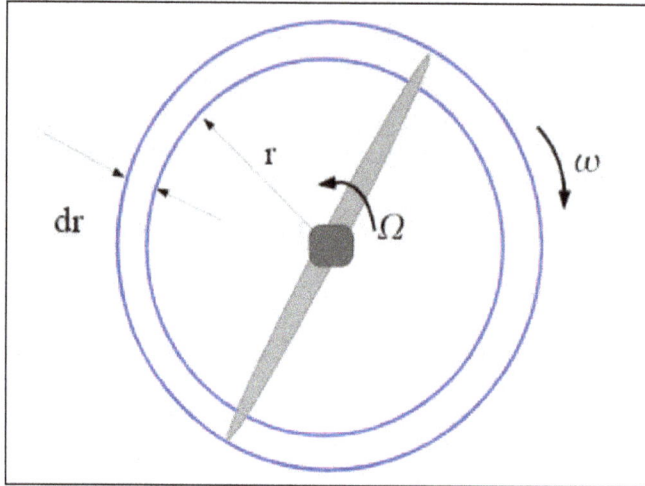

Rotating Annular Stream Tube.

Since each of the blade elements has a different rotational speed and geometric characteristics they will experience a slightly different flow. So blade element theory involves dividing up the blade into a sufficient number (usually between ten and twenty) of elements and calculating the flow at each one. Overall performance characteristics of the blade are then determined by numerical integration along the blade span.

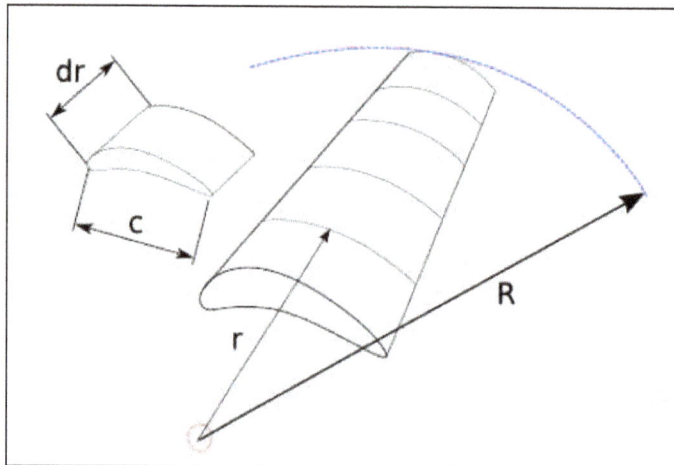

The Blade Element Model.

Lift and drag coefficient data are available for a variety of airfoils from wind tunnel data. Since most wind tunnel testing is done with the aerofoil stationary, the relative velocity over the airfoil is used in order to relate the flow over the moving airfoil with the stationary test.

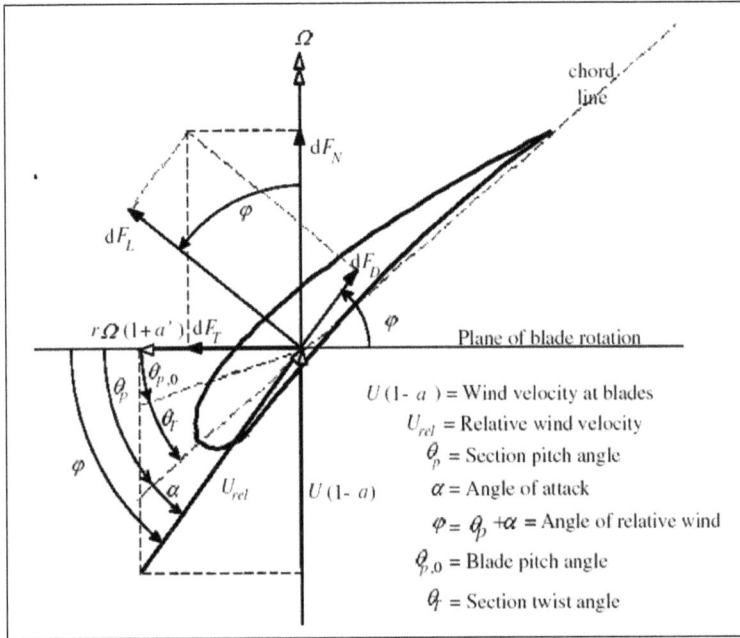

Blade Geometry for the analysis of a HAWT Rotor.

Examining figure the following equations can be derived immediately:

$$U_{rel} = \frac{U_\infty (1-a)}{\sin(\varphi)}$$

$$\tan(\varphi) = \frac{U_\infty (1-a)}{\Omega r (1-a')} = \frac{(1-a)}{(1+a')\lambda_r}$$

$$dF_L = C_L \frac{1}{2} \rho U_{rel}^2 c dr$$

$$dF_D = C_D \frac{1}{2} \rho U_{rel}^2 c dr$$

$$dL = dF_L \sin\varphi - dF_D \cos\varphi$$

$$dT = dF_L \cos\varphi + dF_D \sin\varphi$$

If the rotor has B number of blades, Equation $U_{rel} = \dfrac{U_\infty (1-a)}{\sin(\varphi)}$, $dF_L = C_L \dfrac{1}{2} \rho U_{rel}^2 c dr$ and $dF_D = C_D \dfrac{1}{2} \rho U_{rel}^2 c dr$ can be rearranged.

$$dL = B \frac{1}{2} \rho U_{rel}^2 \left(C_L \sin\varphi - C_D \cos\varphi \right) c dr$$

$$dT = B\frac{1}{2}\rho U_{rel}^2 \left(C_L \cos\varphi + C_D \sin\varphi\right)cdr$$

The elemental torque can be written as dQ = rdL which gives Equation.

$$dQ = B\frac{1}{2}\rho U_{rel}^2 \left(C_L \sin\varphi - C_D \cos\varphi\right)crdr$$

Also Equation can be derived by examining the above figure:

$$U_{rel} = \frac{U_\infty(1-a)}{\sin\varphi}$$

The solidity ratio can be defined as:

$$\sigma = \frac{Bc}{2\pi r}$$

Finally, the general form of elemental torque and thrust equations becomes:

$$dQ = \sigma\pi\rho\frac{U_\infty^2(1-a)^2}{\sin^2\varphi}\left(C_L \sin\varphi - C_D \cos\varphi\right)r^2 dr$$

$$dT = \sigma\pi\rho\frac{U_\infty^2(1-a)^2}{\sin^2\varphi}\left(C_L \cos\varphi - C_D \sin\varphi\right)rdr$$

Equation (mentioned just above) defines the normal force (thrust) and the tangential force (torque) on annular rotor section respectively.

Blade Element Momentum (BEM) Theory

As it is stated before BEM theory refers to the determination of a wind turbine blade performance by combining the equations of general momentum theory and blade element theory, so Equation $dT = 2\rho u\left(u - U_\infty\right)dA = 4\pi\rho U_\infty^2 a(1-a)rdr$ and $dT = \sigma\pi\rho\frac{U_\infty^2(1-a)^2}{\sin^2\varphi}\left(C_L \cos\varphi - C_D \sin\varphi\right)rdr$ can be equated to obtain the following expression.

$$\frac{a}{(1-a)} = (\sigma C_L)\frac{\cos\varphi}{4\sin^2\varphi}\left[1 + (C_D / C_L)\tan\varphi\right]$$

Also equating Equation

$$dQ = 4\pi\rho U_\infty \Omega a'(1-a)r^3 dr$$

and $dQ = \sigma\pi\rho \dfrac{U_\infty^2(1-a)^2}{\sin^2\varphi}(C_L\sin\varphi - C_D\cos\varphi)r^2 dr$ in the same manner gives:

$$\frac{a'}{(1-a)} = \frac{(\sigma C_L)}{4\lambda_r \sin\varphi}\left[1-(C_D/C_L)\cot\varphi\right]$$

By rearranging Equation $\dfrac{a'}{(1-a)} = \dfrac{(\sigma C_L)}{4\lambda_r \sin\varphi}\left[1-(C_D/C_L)\cot\varphi\right]$ and combining it with

Equation $\tan(\varphi) = \dfrac{U_\infty(1-a)}{\Omega r(1-a')} = \dfrac{(1-a)}{(1+a')\lambda_r}$

So,

$$\frac{a'}{(1-a)} = \frac{(\sigma C_L)}{4\cos\varphi}\left[1-(C_D/C_L)\cot\varphi\right]$$

can be written. In order to calculate the induction factors a and a', C_D can be set to zero. Thus the induction factors can be determined independently from airfoil characteristics. Subsequently, the above equations can be rewritten as:

$$\frac{a}{(1-a)} = (\sigma C_L)\frac{\cos\varphi}{4\sin^2\varphi}$$

$$\frac{a'}{(1-a)} = \frac{(\sigma C_L)}{4\lambda_r \sin\varphi}$$

$$\frac{a'}{(1+a')} = \frac{(\sigma C_L)}{4\sin\varphi}$$

Finally, by rearranging above three equations and solving it for a and a', the following useful analytical relationships can be obtained.

$$a = \frac{1}{\left[1+\left[4\sin^2\varphi/(\sigma C_L)\cos\varphi\right]\right]}$$

$$a' = \frac{1}{\left[\left[4\cos\varphi/(\sigma C_L)\right]-1\right]}$$

$$a/a' = \lambda_r/\tan\varphi$$

$$C_L = \frac{4\sin\varphi}{\sigma}\frac{(\cos\varphi - \lambda_r\sin\varphi)}{(\sin\varphi + \lambda_r\cos\varphi)}$$

The total power of the rotor can be calculated by integrating the power of each differential annular element from the radius of the hub to the radius of the rotor.

$$P = \int_{r_h}^{R} dP = \int_{r_h}^{R} \Omega dQ$$

And rewriting the power coefficient given in Equation $C_P = \dfrac{2P}{\rho U_\infty^3 \pi R^2}$ using Equation $P = \int_{r_h}^{R} dP = \int_{r_h}^{R} \Omega dQ$ gives:

$$C_P = \frac{P}{1/2\,\rho U_\infty^3 A} = \frac{\int_{r_h}^{R} \Omega dQ}{1/2\,\rho U_\infty^3 \pi R^2}$$

Using equation,

$$dQ = \sigma \pi \rho \frac{U_\infty^2 (1-a)^2}{\sin^2 \varphi}\left(C_L \sin\varphi - C_D \cos\varphi\right) r^2 dr \,, \quad \frac{a}{(1-a)} = (\sigma C_L)\frac{\cos\varphi}{4\sin^2 \varphi}$$

and $a/a' = \lambda_r / \tan\varphi$ the power coefficient relation can be rearranged as:

$$C_p = \frac{8}{\lambda^2} \int_{\lambda_h}^{\lambda} \lambda_r^3 a'(1-a)\left[1-(C_D/C_L)\cot\varphi\right] d\lambda_r$$

Tip Losses

At the tip of the turbine blade losses are introduced. The ratio of the average value of tip loss factor to that at a blade position is given in figure. As it is shown in the figure only near the tip the ratio begins to fall to zero so it is called 'the tip-loss factor'.

With uniform circulation the azimuthal average value of a is also radially uniform but that implies a discontinuity of axial velocity at the wake boundary with a corresponding discontinuity in pressure. Whereas such discontinuities are acceptable in the idealized actuator disc situation they will not occur in practice with a finite number of blades.

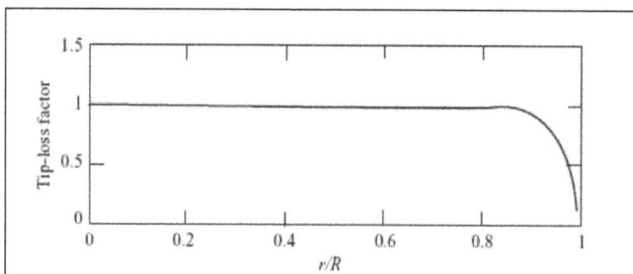

Span-wise Variation of the Tip-loss Factor for a Blade with Uniform Circulation.

The losses at the blade tips can be accounted for in BEM theory by means of a correction factor, f which varies from 0 to 1 and characterizes the reduction in forces along the blade. An approximate method of estimating the effect of tip losses has been given by L. Prandtl and the expression obtained by Prandtl for tip-loss factor is given by Equation:

$$f = \frac{2}{\pi} \cos^{-1} \left\{ \exp \left[\frac{-(B/2)\left[1-\dfrac{r}{R}\right]}{\left(\dfrac{r}{R}\right)\sin\varphi} \right] \right\}$$

The application of this equation for the losses at the blade tips is to provide an approximate correction to the system of equations for predicting rotor performance and blade design. Carrying the tip-loss factor through the calculations, the changes will be as following:

$$dQ = 4f\pi\rho U_\infty \Omega a'(1-a)r^3 dr$$

$$dT = 4f\pi\rho U_\infty^2 a(1-a)r dr$$

$$\frac{a'}{1-a} = \frac{\sigma C_L}{4f\lambda_r \sin\varphi}$$

$$\frac{a'}{1-a'} = \frac{\sigma C_L}{4f\cos\varphi}$$

$$C_L = \frac{4f\sin\varphi}{\sigma} \frac{\left(\cos\varphi - \lambda_{r\sin\varphi}\right)}{\left(\sin\varphi + \lambda_{r\cos\varphi}\right)}$$

$$a = \frac{1}{1 + \left[\dfrac{4f\sin^2\varphi}{(\sigma C_L)\cos\varphi}\right]}$$

$$a' = \frac{1}{\left[\dfrac{4f\cos\varphi}{(\sigma C_L)} - 1\right]}$$

$$C_P = \frac{8}{\lambda^2} \int_{\lambda_h}^{\lambda} f\lambda_r^3 a'(1-a)\left[1 - \frac{C_D}{C_L}\tan\beta\right] d\lambda_r$$

The results for the span-wise variation of power extraction in the presence of tip-loss for a blade with uniform circulation on a three-bladed HAWT operating at a tip speed ratio of 6 is shown in figure and it clearly demonstrates the effects of tip-loss.

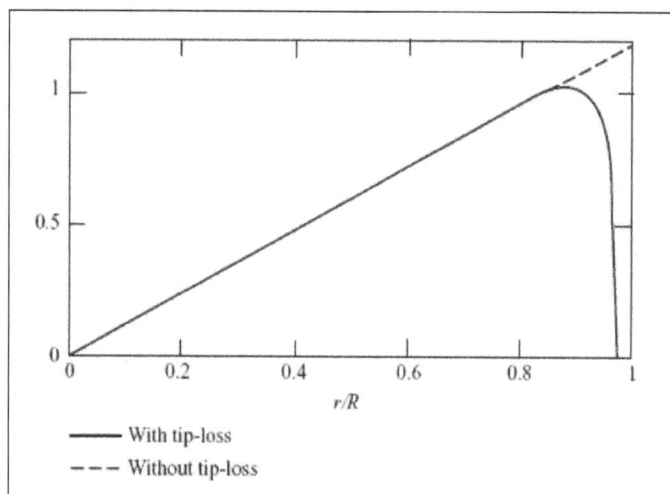

Span-wise Variation of Power Extraction.

Accordingly, for a selected airfoil type and a specified tip-speed ratio with blade length, the blade geometry can be designed for optimum rotor. And using these geometric parameters determined the aerodynamic performance of the rotor can be analyzed.

Blade Design Procedure

The aerodynamic design of optimum rotor blades from a known airfoil type means determining the geometric parameters (such as chord length and twist angle distribution along the blade span) for a certain tip-speed ratio at which the power coefficient of the rotor is maximum. For this reason firstly the change of the power coefficient of the rotor with respect to tip-speed ratio should be figured out in order to determine the design tip-speed ratio λ_d where the rotor has a maximum power coefficient. The blade design parameters will then be according to this design tip-speed ratio. Examining the plots between relative wind angle and local tip-speed ratio for a wide range of glide ratios gives us a unique relationship when the maximum elemental power coefficient is considered. And this relationship can be found to be nearly independent of glide ratio and tip-loss factor. Therefore a general relationship can be obtained between optimum relative wind angle and local tip-speed ratio which will be applicable for any airfoil type.

$$\frac{\partial}{\partial \varphi}\left\{\sin^2 \varphi\left(\cos \varphi - \lambda_r \sin \varphi\right)\left(\sin \varphi + \lambda_r \cos \varphi\right)\right\} = 0$$

Equation above reveals after some algebra:

$$\varphi_{opt} = \left(2/3\right)\tan^{-1}\left(1/\lambda_r\right)$$

Having found the solution of determining the optimum relative wind angle for a certain local tip-speed ratio, the rest is nothing but to apply the equations from 80 to 83 derived from the blade-element momentum theory and modified including the tip loss factor. Dividing the blade length into N elements, the local tip-speed ratio for each blade element can then be calculated as:

$$\lambda_{r,i} = \lambda\left(r_i / R\right)$$

Then rewriting Equation for each blade element gives:

$$\varphi_{opt,i} = \left(2/3\right)\tan^{-1}\left(1/\lambda_{r,i}\right)$$

In addition the tip loss correction factor for each element can be calculated as:

$$f_i = \frac{2}{\pi}\cos^{-1}\left\{\exp\left[\frac{-\left(B/2\right)\left[1-\left(\dfrac{r_i}{R}\right)\right]}{\left(\dfrac{r_i}{R}\right)\sin\varphi_{opt,i}}\right]\right\}$$

The local chord-length for each blade element can then be determined using the following expression:

$$c_i = \frac{8\pi r_i F_i \sin\varphi_{opt,i}}{BC_{L,design}}\frac{\left(\cos\varphi_{opt,i} - \lambda_{r,i}\sin\varphi_{opt,i}\right)}{\left(\sin\varphi_{opt,i} + \lambda_{r,i}\cos\varphi_{opt,i}\right)}$$

where, $C_{L_{design}}$ is chosen such that the glide ratio is minimum at each blade element. Also the twist distribution can easily be calculated by Equation:

$$\theta_i = \varphi_{opt,i} - \alpha_{design}$$

where, α_{design} is again the design angle of attack at which $C_{L_{design}}$ is obtained. Now chord length and twist distribution along the blade span are known and in this case lift coefficient and angle of attack have to be determined from the known blade geometry parameters. This requires an iterative solution in which for each blade element the axial and angular induction factors are firstly taken as the values equal to the corresponding designed blade elements. Then the actual induction factors are determined within an acceptable tolerance of the previous guesses during iteration.

The kinetic energy extracted from the wind is influenced by the geometry of the rotor blades. Determining the aerodynamically optimum blade shape, or the best possible approximation to it, is one of the main tasks of the wind turbine designer. Accordingly this chapter sets out the basis of the aerodynamics of HAWTs and the design methods based on these theories to find the best possible design compromise for the geometric shape of the rotor which can only be achieved in an iterative process.

Performance analysis of HAWT rotors has been performed using several methods. In between these methods BEM model is mainly employed as a tool of performance analysis due to its simplicity and readily implementation. Most wind turbine design codes are based on this method. Accordingly the chapter explains the aerodynamics of HAWTs based on a step-by-step approach starting from the simple actuator disk model to more complicated and accurate BEM method.

The basic of BEM method assumes that the blade can be analyzed as a number of independent elements in span-wise direction. The induced velocity at each element is determined by performing the momentum balance for an annular control volume containing the blade element. Then the aerodynamic forces on each element are calculated using the lift and drag coefficient from the empirical two-dimensional wind tunnel test data at the geometric angle of attack of the blade element relative to the local flow velocity. BEM theory-based methods have aspects by reasonable tool for designer, but they are not suitable for accurate estimation of the wake effects, the complex flow such as three-dimensional flow or dynamic stall because of the assumptions being made.

References

- Aerodynamics, entry: newworldencyclopedia.org, Retrieved 25 January, 2019

- Angle-of-attack, theory: aviation-history.com, Retrieved 18 March, 2019

- Drag-coefficient- 627: engineeringtoolbox.com, Retrieved 05 June, 2019

- Forces-acting-on-the-aircraft-lift: flightliteracy.com, Retrieved 17 May, 2019

- Thrust-reversal: aerospaceengineeringblog.com, Retrieved 12 July, 2019

- Helicopter-aerodynamics-of-flight: aircraftsystemstech.com, Retrieved 18 April, 2019

- Launcher-aerodynamics-a-suitable-investigation-approach-at-phase-a-design-level, advances-in - some-hypersonic-vehicles-technologies: intechopen.com, Retrieved 14 May, 2019

- Aerodynamics-of-wind-turbines, fundamental-and-advanced-topics-in-wind-power: intechopen. com, Retrieved 19 February, 2019

Aerodynamics: Fundamental Principles and Equations

Some of the fundamental principles and equations of aerodynamics include conservation of energy, compressible and incompressible flow, aerodynamic force and circulation, velocity equation, momentum equation, etc. All these principles and equations of aerodynamics have been carefully analyzed in this chapter.

Motion

The continuous change in position of a body with respect to time and relative to the reference point or observer is called motion. Here the reference point can be considered as stationary or motionless objects surrounding the body.

Example of Motion:

Consider sun as reference point or observer the planets orbiting the sun are said to be in motion. The position of a body is not changing with time and relative to the reference point then the body is said to be at rest or motionless. Chairs, tables, Television are some examples of objects which are at rest. Rest and motion are relative terms for example consider a person in a moving bus he is at rest with respect to another person in the moving bus, but he is in motion with respect person out side the bus.

Uniform Motion

When a body moves equal distance in equal intervals of time then the body is said to be in uniform motion.

Example:

A train travels 10 Km distance at first 10 min (minutes) and the next 10 min (minutes) the train covers the same 10 Km distance then the train is said to be at uniform motion. Speed of light in vacuum is an example of uniform motion.

Non-uniform Motion

When a body moves unequal distance in equal intervals of time then the body is said to be in non-uniform motion.

Example:

A train travels 10 Km distance in first 10 min (minutes) and the next 10 min (minutes) the train travels 20 Km then the train is said to be in non uniform motion because after the first 10 min the speed of train gets increased so the train is not traveling at constant speed.

Periodic Motion

When the motion is repeated for a fixed period of time then the motion is said to be periodic motion. For example planets orbiting the sun is called periodic motion because planets are moving in the same path continuously.

Oscillatory Motion

The oscillatory motion is a specific type of periodic motion in which a body moves back and forth i.e. either up and down or front and back for a fixed period of time from the initial point.

Airspeed

Speed is the distance traveled per unit of time. It is how fast an object is moving. Speed is the scalar quantity that is the magnitude of the velocity vector. It doesn't have a direction. Higher speed means an object is moving faster. Lower speed means it is moving slower. If it isn't moving at all, it has zero speed.

The most common way to calculate the constant velocity of an object moving in a straight line is the formula:

$$r = d / t$$

where,

r is the rate, or speed (sometimes denoted as v, for velocity),

d is the distance moved,

t is the time it takes to complete the movement.

This equation gives the average speed of an object over an interval of time.

The object may have been going faster or slower at different points during the time interval, but we see here its average speed.

The instantaneous speed is the limit of the average speed as the time interval approaches zero. When you look at a speedometer in a car, you are seeing the instantaneous speed. While you may have been going 60 miles per hour for a moment, your average rate of speed for ten minutes might be far more or far less.

Units for Speed

The SI units for speed are m/s (meters per second). In everyday usage, kilometers per hour or miles per hour are the common units of speed. At sea, knots (or nautical miles) per hour is a common speed.

Conversions for Unit of Speed

	km/h	mph	knot	ft/s
1 m/s =	3.6	2.236936	1.943844	3.280840

Rotational Speed and Tangential Speed

Rotational speed, or angular speed, is the number of revolutions over a unit of time for an object traveling in a circular path. Revolutions per minute (rpm) are a common unit. But how far from the axis an object is its radial distance as it revolves determines its tangential speed, which is the linear speed of an object on a circular path?

At one rpm, a point that is at the edge of a record disk is covering more distance in a second than a point closer to the center. At the center, the tangential speed is zero. Your tangential speed is proportional to the radial distance times the rate of rotation.

Tangential speed = radial distance × rotational speed.

Airspeed is the velocity of an airplane relative to the air mass through which it is flying. In simple terms, it's the result of thrust impeded by drag. How fast an airplane can go in level flight depends on the amount of drag and the amount of horsepower. The sleek Grob 115, for example, has less drag per horsepower than a Pitts Special, which has an extra set of wings and their associated struts and wire bracing.

All airspeeds, whether flat-out or economy cruise, are the result of drag equaling the selected amount of power. They are determined by the balance of an aircraft's pitch attitude and power setting, and each result in specific aircraft performance or denote an operating limitation.

Reading Speed

While you determine airspeed with pitch and power, you read it on the airspeed indicator (ASI), which may be marked in miles per hour (mph), nautical miles per hour (knots), or both. While this may seem as simple as reading the number to which the needle points, there's more to it because there are three types of airspeed.

Calibrated Airspeed

Calibrated airspeed (CAS) is indicated airspeed corrected for instrument and position error.

When flying at sea level under International Standard Atmosphere conditions (15 °C, 1013 hPa, 0% humidity) calibrated airspeed is the same as equivalent airspeed (EAS) and true airspeed (TAS). If there is no wind it is also the same as ground speed (GS). Under any other conditions, CAS may differ from the aircraft's TAS and GS.

Calibrated airspeed in knots is usually abbreviated as *KCAS*, while indicated airspeed is abbreviated as *KIAS*.

In some applications, notably British usage, the expression *rectified airspeed* is used instead of calibrated airspeed.

Practical Applications of CAS

CAS has two primary applications in aviation:

- For navigation, CAS is traditionally calculated as one of the steps between indicated airspeed and true airspeed.

- For aircraft control, CAS (and EAS) are the primary reference points, since they describe the dynamic pressure acting on aircraft surfaces regardless of density, altitude, wind, and other conditions. EAS is used as a reference by aircraft designers, but EAS cannot be displayed correctly at varying altitudes by a simple (single capsule) airspeed indicator. CAS is therefore a standard for calibrating the airspeed indicator such that CAS equals EAS at sea level pressure and approximates EAS at higher altitudes.

With the widespread use of GPS and other advanced navigation systems in cockpits, the first application is rapidly decreasing in importance – pilots are able to read groundspeed (and often true airspeed) directly, without calculating calibrated airspeed as an intermediate step. The second application remains critical, however – for example, at the same weight, an aircraft will rotate and climb at approximately the same calibrated airspeed at any elevation, even though the true airspeed and groundspeed may differ significantly. These V speeds are usually given as IAS rather than CAS, so that a pilot can read them directly from the airspeed indicator.

Calculation from Impact Pressure

Since the airspeed indicator capsule responds to impact pressure, CAS is defined as a function of impact pressure alone. Static pressure and temperature appear as fixed coefficients defined by convention as standard sea level values. It so happens that the speed of sound is a direct function of temperature, so instead of a standard temperature, we can define a standard speed of sound.

For subsonic speeds, CAS is calculated as:

$$CAS = a_0 \sqrt{5\left[\left(\frac{q_c}{P_0}+1\right)^{\frac{2}{7}} -1\right]}$$

Where:

- q_c = impact pressure,
- P_0 = standard pressure at sea level,
- a_0 is the standard speed of sound at 15 °C.

For supersonic airspeeds, where a normal shock forms in front of the pitot probe, the Rayleigh formula applies:

$$CAS = a_0 \left[\left(\frac{q_c}{P_0}+1\right) \times \left(7\left(\frac{CAS}{a_0}\right)^2 -1\right)^{2.5} /\left(6^{2.5} \times 1.2^{3.5}\right)\right]^{(1/7)}$$

The supersonic formula must be solved iteratively, by assuming an initial value for CAS equal to a_0.

These formulae work in any units provided the appropriate values for P_0 and a_0 are selected. For example, P_0 = 1013.25 hPa, a_0 = 1,225 km/h (661.45 kn). The ratio of specific heats for air is assumed to be 1.4.

These formulae can then be used to calibrate an airspeed indicator when impact pressure (q_c) is measured using a water manometer or accurate pressure gauge. If using a water manometer to measure millimeters of water the reference pressure (P_0) may be entered as 10333 mm H_2O.

At higher altitudes CAS can be corrected for compressibility error to give equivalent airspeed (EAS). In practice compressibility error is negligible below about 3,000 m (10,000 ft) and 370 km/h (200 kn).

Indicated Airspeed

Indicated airspeed (IAS) is the airspeed read directly from the airspeed indicator (ASI) on an aircraft, driven by the pitot-static system. It uses the difference between total pressure and static pressure, provided by the system, to either mechanically or electronically measure dynamic pressure. The dynamic pressure includes terms for both density and airspeed. Since the airspeed indicator cannot know the density, it is by design calibrated to assume the sea level standard atmospheric density when calculating airspeed. Since the actual density will vary considerably from this assumed value as the aircraft changes altitude, IAS varies considerably from true airspeed (TAS), the relative velocity between the aircraft and the surrounding air mass. Calibrated airspeed (CAS) is the IAS corrected for instrument and position error.

An aircraft's indicated airspeed in knots is typically abbreviated *KIAS* for "Knots-Indicated Air Speed" (vs. *KCAS* for calibrated airspeed and *KTAS* for true airspeed).

The IAS is an important value for the pilot because it is the indicated speeds which are specified in the aircraft flight manual for such important performance values as the stall speed. These speeds, in true airspeed terms, vary considerably depending upon density altitude. However, at typical civilian operating speeds, the aircraft's aerodynamic structure responds to dynamic pressure alone, and the aircraft will perform the same when at the same dynamic pressure. Since it is this same dynamic pressure that drives the airspeed indicator, an aircraft will always, for example, stall at the published *indicated* airspeed (for the current configuration) regardless of density, altitude or true airspeed.

Furthermore, the IAS is specified in some regulations, and by air traffic control when directing pilots, since the airspeed indicator displays that speed (by definition) and it is the pilot's primary airspeed reference when operating below transonic or supersonic speeds.

Calculation

Indicated airspeed measured by pitot-tube can be expressed by following equation delivered from Bernoulli's equation.

$$u = \sqrt{\frac{2(p_t - p_s)}{\rho}}$$

The above equation applies only to conditions that can be treated as incompressible. Liquids are treated as incompressible under almost all conditions. Gases under certain conditions can be approximated as incompressible. The compression effects can be corrected by use of Poisson constant. This compensation corresponds to Equivalent airspeed.

$$u = \sqrt{\frac{2\gamma}{\gamma-1} \frac{p_s}{\rho} \left[\left(\frac{p_t}{p_s} \right)^{\frac{\gamma-1}{\gamma}} - 1 \right]}$$

Where:

- u is indicated airspeed in m/s.

- p_t is stagnation or total pressure in pascals.

- p_s is static pressure in pascals.

- ρ is fluid density in kg/m^3.

- γ is the Poisson constant of a gas at a constant pressure to heat at a constant volume (1.4 for air).

IAS vs. CAS

The IAS is not the actual speed through the air even when the aircraft is at sea level under International Standard Atmosphere conditions (15 °C, 1013 hPa, 0% humidity). The IAS needs to be corrected for known instrument and position errors to show true airspeed under those specific atmospheric conditions, and this is the CAS (Calibrated Airspeed). Despite this the pilot's primary airspeed reference, the ASI, shows IAS (by definition). The relationship between CAS and IAS is known and documented for each aircraft type and model.

IAS and V Speeds

The aircraft's pilot manual usually gives critical V speeds as IAS, those speeds indicated by the airspeed indicator. This is because the aircraft behaves similarly at the same IAS no matter what the TAS is: E.g. A pilot landing at a hot and high airfield will use the same IAS to fly the aircraft at the correct approach and landing speeds as he would when landing at a cold sea level airfield even though the TAS must differ considerably between the two landings.

Whereas IAS can be reliably used for monitoring critical speeds well below the speed of sound this is not so at higher speeds. An example: Because (1) the compressibility of air changes considerably approaching the speed of sound, and (2) the speed of sound varies considerably with temperature and therefore altitude; the maximum speed at which an aircraft structure is safe, the never exceed speed (abbreviated V_{NE}), is specified at several differing altitudes in faster aircraft's operating manuals, as shown in the sample table below:

Diving below	IAS mph	IAS km/h
30,000 ft (9,100 m)	370	595
25,000 ft (7,600 m)	410	660
20,000 ft (6,100 m)	450	725
15,000 ft (4,600 m)	490	790
10,000 ft (3,000 m)	540	870

IAS and Navigation

For navigation, it is necessary to convert IAS to TAS and ground speed (GS) using the following method:

- Correct IAS to calibrated airspeed (CAS) using an aircraft-specific correction table.

- Correct CAS to true airspeed (TAS) by using Outside Air Temperature (OAT), Pressure-altitude and CAS on an E6B flight computer or equivalent functionality on most gpss.

- Convert TAS to ground speed (GS) by allowing for the effect of wind.

With the advent of Doppler radar navigation and, more recently, GPS receivers, with other advanced navigation equipment that allows pilots to read ground speed directly, the TAS calculation in-flight is becoming unnecessary for the purposes of navigation estimations.

TAS is the primary method to determine aircraft's cruise performance in manufacturer's specs, speed comparisons and pilot reports.

True Airspeed

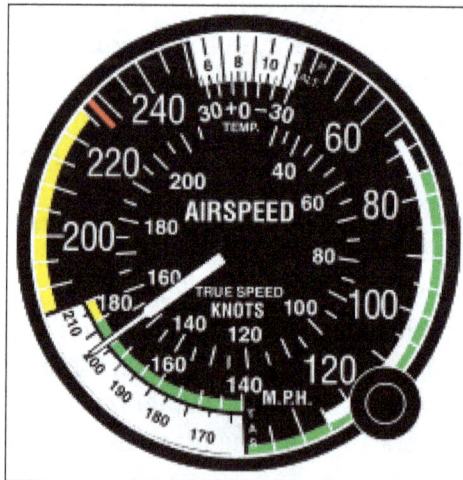

A mechanical true airspeed indicator for an airplane.

The pilot sets the pressure altitude and air temperature in the top window using the knob; the needle indicates true airspeed in the lower left window. Here the speed is displayed both in knots (kn) and miles per hour (mph).

The true airspeed (TAS; also KTAS, for *knots true airspeed*) of an aircraft is the speed of the aircraft relative to the airmass in which it is flying. The true airspeed is important information for accurate navigation of an aircraft. Traditionally it is measured using an analogue TAS indicator, but as the Global Positioning System has become available for civilian use, the importance of such analogue instruments has decreased. Since indicated airspeed is a better indicator of power used and lift available, True airspeed is not used for controlling the aircraft during taxiing, takeoff, climb, descent, approach or landing; for these purposes the Indicated airspeed – IAS or KIAS (knots indicated airspeed) – is used. However, since indicated airspeed only shows true speed through the air at standard sea level pressure and temperature, a TAS meter is necessary for navigation purposes at cruising altitude in less dense air. The IAS meter reads very nearly the TAS at lower altitude and at lower speed. On jet airliners the TAS meter is usually hidden at speeds below 200 knots (370 km/h). Neither provides for accurate speed over the ground, since surface winds or winds aloft are not taken into account.

Performance

TAS is the true measure of aircraft performance in cruise, thus it is the speed listed in aircraft specifications, manuals, performance comparisons, pilot reports, and every situation when cruise or endurance performance needs to be measured. It is the speed normally listed on the flight plan, also used in flight planning, before considering the effects of wind.

Airspeed Sensing Errors

The airspeed indicator (ASI), driven by ram air into a Pitot tube and still air into a barometric static port, and shows what is called indicated airspeed (IAS). The differential pressure is affected by air density. The ratio between the two measurements is temperature-dependent and pressure-dependent, according to the ideal gas law.

At sea level in the International Standard Atmosphere (ISA) and at low speeds where air compressibility is negligible (i.e., assuming a constant air density), IAS corresponds to TAS. When the air density or temperature around the aircraft differs from standard sea level conditions, IAS will no longer correspond to TAS, thus it will no longer reflect aircraft performance. The ASI will indicate less than TAS when the air density decreases due to a change in altitude or air temperature. For this reason, TAS cannot be measured directly. In flight, it can be calculated either by using an E6B flight calculator or its equivalent.

For low speeds, the data required are static air temperature, pressure altitude and IAS (or CAS for more precision). Above approximately 100 knots (190 km/h), the compressibility error rises significantly and TAS must be calculated by the Mach speed. Mach incorporates the above data including the compressibility factor. Modern aircraft instrumentation use an Air Data Computer to perform this calculation in real time and display the TAS reading directly on the EFIS.

Since temperature variations are of a smaller influence, the ASI error can be roughly estimated as indicating about 2% less than TAS per 1,000 feet (300 m) of altitude above sea level. For example, an aircraft flying at 15,000 feet (4,600 m) in the international standard atmosphere with an IAS of 100 knots (190 km/h), is actually flying at 126 knots (233 km/h) TAS.

Use in Navigation Calculations

To maintain a desired ground track while flying in the moving airmass, the pilot of an aircraft must use knowledge of wind speed, wind direction, and true air speed to determine the required heading.

Calculating True Airspeed

Low-speed Flight

At low speeds and altitudes, IAS and CAS are close to equivalent airspeed (EAS). TAS can be calculated as a function of EAS and air density:

$$TAS = EAS\sqrt{\frac{\rho_0}{\rho}},$$

where,

TAS is true airspeed,

EAS is equivalent airspeed,

ρ_0 is the air density at sea level in the International Standard Atmosphere (15 °C and 1013.25 hectopascals, corresponding to a density of 1.225 kg/m³),

ρ is the density of the air in which the aircraft is flying.

High-speed Flight

TAS can be calculated as a function of Mach number and static air temperature:

$$TAS = a_0 M\sqrt{\frac{T}{T_0}},$$

where,

a_0 is the speed of sound at standard sea level (661.47 knots (1,225.04 km/h; 340.29 m/s)),

M is Mach number,

T is static air temperature in kelvins,

T_0 is the temperature at standard sea level (288.15 K).

For manual calculation of TAS in knots, where Mach number and static air temperature are known, the expression may be simplified to:

$$TAS = 39M\sqrt{T}$$

Combining the above with the expression for Mach number gives an expression for TAS as a function of impact pressure, static pressure and static air temperature (valid for subsonic flow):

$$\text{TAS} = a_0 \sqrt{\frac{5T}{T_0}\left[\left(\frac{q_c}{P}+1\right)^{\frac{2}{7}}-1\right]},$$

where,

q_c is impact pressure,

P is static pressure.

Electronic flight instrument systems (EFIS) contain an air data computer with inputs of impact pressure, static pressure and total air temperature. In order to compute TAS, the air data computer must convert total air temperature to static air temperature. This is also a function of Mach number:

$$T = \frac{T_t}{1+0.2M^2},$$

where,

T_t = total air temperature.

In simple aircraft, without an air data computer or Machmeter, true airspeed can be calculated as a function of calibrated airspeed and local air density (or static air temperature and pressure altitude, which determine density). Some airspeed indicators incorporate a slide rule mechanism to perform this calculation. Otherwise, it can be performed using this applet or a device such as the E6B (a handheld circular slide rule).

Flying Colors

The different airspeeds that either limit or result in specific aircraft performance are known as V speeds—V for velocity. There are a multitude of them, with many, such as VWW, the maximum speed at which you can operate the windshield wipers (honest), applying only to certain airplanes.

On airplanes with a maximum gross weight of less than 12,500 pounds and certificated after 1945, some of the more important V speeds are color-coded on the ASI. This enables pilots to quickly determine how their aircraft velocity conforms to certain airspeed limitations, and whether they are at a safe speed for their current phase of flight or need to either increase or decrease their speed.

The white arc covers the speeds at which the aircraft can be flown with its flaps fully extended. VSO is at the low end of the white arc. This is the speed at which the airplane will stall in straight flight (turns increase the aircraft's load factor, and thereby its stall speed) when at maximum gross weight with the power at idle, fully extended flaps, landing gear down (if so equipped), and with its center of gravity (CG) at its aft limit. VSO is an important speed to monitor, especially when landing.

VFE, the maximum velocity at which the airplane can be flown with its flaps fully extended, is the high-speed limit of the white arc. Flying at speeds greater than VFE with full flaps can result in damage, perhaps to the point of losing one or both flaps. Not a good thing. A number of airplanes, however, do allow the use of approach flaps, usually around 10 degrees, at speeds higher than VFE. The POH will give the specific details.

The green arc spans the aircraft's normal operating speed range. It starts with VS, the velocity at which the airplane will stall in straight flight when at maximum gross weight, the power at idle, the flaps and gear retracted, and aft CG.

The green arc terminates at VNO, the maximum normal operating velocity or maximum structural cruising speed. The formula for calculating VNO is somewhat complex. But one of the formula's factors is the airplane's ability to withstand a specified vertical gust (30 feet per second for planes certificated before August 1969 and 50 feet per second after this date) and not exceed its maximum load limit. It's important to remember that VNO is a certification value. Only maneuvering speed (VA), which will be addressed shortly, will protect you from harm in turbulence.

VNO is also the start of the yellow arc, often called the caution range. Flight in this speed range should only be considered when the air is glass-smooth because the slightest burble of air may cause you to exceed the aircraft's maximum load factor.

The yellow arc terminates at the red line—VNE—the velocity that should never be exceeded. VNE is 90 percent or less of the demonstrated dive velocity (VD), a calculated value and the speed at which a test pilot flew the plane with no vibration or buffeting severe enough to result in structural damage. Don't think there's a 10-percent safety buffer past VNE. A baby's breath will cause the aircraft to exceed its limit load factor, and structural damage will result.

Tree Speeds

At some point during their training, all pilots learn about VX and VY. And in many cases, because both are climb speeds, they spend the rest of their flying careers trying to remember which is which.

VX is the velocity that will give you the most altitude in the shortest distance. It's your aircraft's best angle of climb speed, the speed to use when there is a sequoia or mountain at the end of the runway. It will take more time to gain altitude at VX because you're flying at a slower speed. But gaining altitude in a short amount of time isn't your goal. Gaining altitude in the shortest horizontal distance, like before you hit that tree or mountain, is.

When faced with such a situation, don't pull the airplane off the ground early and try to force it over the obstacle. Airplanes fly at slower speeds in ground effect, and when you climb out of it, you may settle back to earth. Not the thing to do when you want to gain a lot of altitude in a short horizontal distance (the distance between the tree and where you start your takeoff).

And don't hold the airplane on the ground past VX, thinking that the extra speed will rocket you up and over the obstacle. Assuming that the acceleration is essentially unaffected, takeoff distance varies as the square of the takeoff velocity. In other words, 10 percent excess airspeed would increase the takeoff distance 21 percent.

If obstacle X is not impeding your takeoff path, VY is your post-takeoff climb speed because it gives the best rate of climb. You'll gain a lot of altitude in a short amount of time—and you'll also cover a lot a ground. VY also provides for better visibility and engine cooling because of its lower pitch angle when compared to VX.

Whether you're going up, coming down, or going from Point A to B, heeding your aircraft's speed limitations and flying at the airspeeds that will result in the desired performance is a key to safety. While speed may be life in certain situations, it can also kill. Use it wisely.

Getting Down

VREF for Light Aircraft

Because stall speed varies with aircraft weight, the pilots of large, heavy airplanes approach and land using reference speeds—VREF—that are based on the stall speed and other factors at the aircraft's landing weight.

Although the differences between takeoff and landing weights are not as great as they are with large airplanes, there's no reason why pilots of light aircraft can't benefit from VREF speeds. VREF gives you a landing speed that provides a margin of safety above stall speed but is not so fast that the plane will float the length of the runway.

Before you modify your landing speeds, consult and follow POH recommendations. Your VREF speeds are based on your aircraft's VSO. You should use the figure that represents the aircraft's landing weight (or as close to it as you can get, erring on the high side). If your VSO is given for maximum gross weight only, you can adjust it for lighter weights with the same formula used to determine VA at less than max gross weight (10 percent reduction in weight, 5 percent reduction in VSO).

It's important to know whether your aircraft's VSO is given in IAS or CAS when determining VREF speeds. CAS should be used when applicable and converted to IAS for practical use because the difference between the two can be quite large.

Unless your POH recommends otherwise, fly the pattern no faster than VFE and no slower than 1.4 times VSO. This keeps your speed up in the pattern and gives you full use of your flaps and a safety margin over stall speed. Maintain this speed until you turn final. Then let your speed decay to 1.3 times VSO once the landing gear and full flaps are deployed. Remember, VSO is only an accurate stall speed in this specific landing configuration.

If the wind is blowing, add one-half the gust factor to your landing speed. If the wind is 10 knots gusting to 20, add half the difference (5 knots) to your speed. Remember, 1.3 VSO gives you a safety margin, but only after all maneuvering is completed and full flaps and gear are down. So use 1.3 VSO on short final only.

Velocity

Velocity is defined as a vector measurement of the rate and direction of motion. Put simply, velocity is the speed at which something moves in a particular direction, such as the speed of a car traveling north on a major freeway, or the speed a rocket travels as it launches into space. The scalar (absolute value) magnitude of the velocity vector is the speed of the motion. In calculus terms, velocity is the first derivative of position with respect to time. You can calculate velocity by using a simple formula that uses rate, distance, and time.

Velocity Formula

The most common way to calculate the constant velocity of an object moving in a straight line is with the formula:

$$r = d / t$$

where,

- r is the rate, or speed (sometimes denoted as v, for velocity),

- d is the distance moved,

- t is the time it takes to complete the movement.

Units of Velocity

The SI (international) units for velocity are m/s (meters per second). But velocity may be expressed in any units of distance per time. Other units include miles per hour (mph), kilometers per hour (kph), and kilometers per second (km/s).

Importance of Velocity

Velocity measures motion starting in one place and heading toward another place. In other words, you use measures of velocity to determine how quickly you (or anything in motion) will arrive at a destination from a given location.

Measures of velocity allow you to (among other things) create timetables for travel. For example, if a train leaves Penn Station in New York at 2 p.m. and you know the velocity at which the train is moving north, you can predict when it will arrive at South Station in Boston.

Relative Velocity

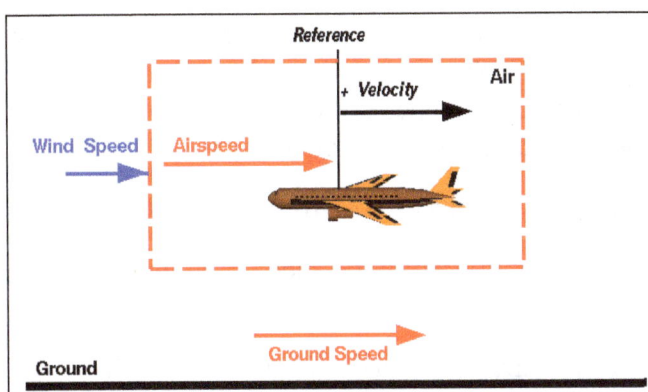

One of the most confusing concepts for young aerodynamicists is the relative velocity between objects. Aerodynamic forces are generated by an object moving through the air. Aerodynamic lift, for instance, depends on the square of the velocity between the object and the air. Things get confusing because not only can the object be moved through the air, but the air itself can move. To properly define the velocity, it is necessary to pick a fixed reference point and measure velocities relative to the fixed point. In this slide, the reference point is fixed to the airplane, but it could just as easily be fixed to the ground.

The important quantity in the generation of lift is the relative velocity between the object and the air. For a reference point picked on the aircraft, the air moves relative to the reference point at the airspeed. The airspeed is a vector quantity and has both a magnitude and a direction. A positive velocity is defined to be toward the tail of the aircraft. The airspeed can be directly measured on the aircraft by use of a pitot tube. For a reference point picked on the aircraft, the ground moves aft at some velocity called the ground speed. Ground speed is also a vector quantity so a comparison with the airspeed must be done according to the rules of vector comparisons.

The air in which the aircraft fly's can move in all three directions. In this figure, we are only considering velocities along the aircraft's flight path and we are neglecting cross winds which occur perpendicular to the flight path but parallel to the ground and updrafts and downdrafts which occur perpendicular to the ground. From the aircraft, we can not directly measure the wind speed, but must compute the wind speed from the ground speed and airspeed. Wind speed is the vector difference between the airspeed and the ground speed.

Wind speed = Airspeed - Ground Speed

On a perfectly still day the wind speed is zero and the airspeed is equal to the ground speed. If the measured airspeed is greater than the observed ground speed, the wind speed is positive.

Suppose we had an airplane which could take off on a windless day at 100 mph (lift off airspeed is 100 mph). Now suppose we had a day in which the wind was blowing 20 mph towards the West. If the airplane takes off going East, it experiences a 20 mph headwind (wind in your face). Since a positive velocity is defined to be toward the tail, a headwind will be a positive wind speed. While the plane is sitting still on the runway, it has a ground speed of 0 and an airspeed of 20 mph.

Wind speed (20) = Airspeed (20) - Ground Speed (0)

At lift off, the airspeed is 100 mph, the wind speed is 20 mph and the ground speed will be 80 mph:

Wind speed (20) = Airspeed (100) - Ground Speed (80)

If the plane took off to the West it would have a 20 mph tail wind (wind at your back). This gives a negative wind speed. At lift off, the airspeed is still 100 mph, the wind speed is -20 mph and the ground speed will now be 120 mph.

Wind speed (-20) = Airspeed (100) - Ground Speed (120)

So the aircraft will have to travel faster (and farther) along the ground to achieve lift off conditions with the wind at it's back.

Aerodynamic Force

Aerodynamic force is the force exerted on a body whenever there is a relative velocity between the body and the air. There are only two basic sources of aerodynamic force: the pressure distribution and the frictional shear stress distribution exerted by the airflow on the body surface. The pressure exerted by the air at a point on the surface acts perpendicular to the surface at that point; and the shear stress, which is due to the frictional action of the air rubbing against the surface, acts tangentially to the surface at that point. The distribution of pressure and shear stress represent a distributed load over the surface. The net aerodynamic force on the body is due to the net imbalance between these distributed loads as they are summed (integrated) over the entire surface.

For purposes of discussion, it is convenient to consider the aerodynamic force on an airfoil. The net resultant aerodynamic force R acting through the center of pressure on the airfoil represents mechanically the same effect as that due to the actual pressure and shear stress loads distributed over the body surface. The velocity of the airflow V_∞ is called the free-stream velocity or the free-stream relative wind. By definition, the component of R perpendicular to the relative wind is the lift, L, and the component of R parallel to the relative wind is the drag D. The orientation of the body with respect to the direction of the free stream is given by the angle of attack, α. The magnitude of the aerodynamic force R is governed by the density and rgr $_\infty$ velocity of the free stream, the size of the body, and the angle of attack.

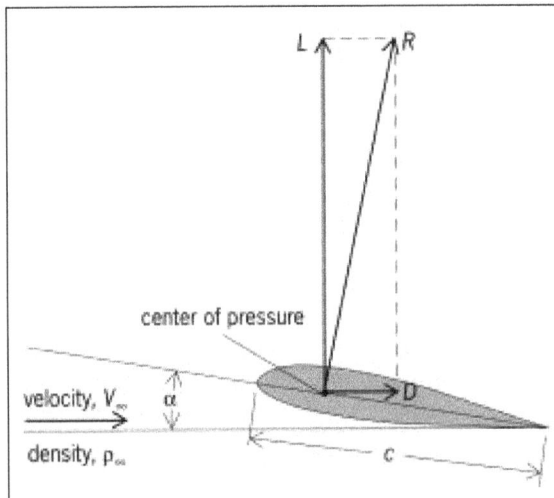

Resultant aerodynamic force (R), and its resolution
into lift (L) and drag (D) components.

An important measure of aerodynamic efficiency is the ratio of lift to drag, L/D. The higher the value of L/D, the more efficient is the lifting action of the body. The value of L/D reaches a maximum, denoted by (L/D)$_{max}$, at a relatively low angle of attack. Beyond a certain angle the lift decreases with increasing α. In this region, the wing is said to be stalled. In the stall region the flow has separated from the top surface of the wing, creating a type of slowly recirculating dead-air region, which decreases the lift and substantially increases the drag.

Aerodynamic Circulation

The word circulation is not meant for a single meaning, in aerodynamics Circulation is a mathematical method of calculating the lift generated by an airfoil using the physics of fluid circulation. It was generated as a method of calculating the lift generated by rotating bodies (tennis balls, golf balls and baseballs). It has little practical application to airplanes, since the air does not physically circulate around the airfoil.

Circulation is the line integral of the velocity field, around a closed curve. Circulation is normally denoted Γ (uppercase gamma). Circulation was first used independently by Frederick Lanchester, Martin Kutta and Nikolai Zhukovsky.

If $\mathbf{V}$ is the fluid velocity on a small element of a defined curve, and $d\mathbf{l}$ is a vector representing the differential length of that small element, the contribution of that differential length to circulation is $d\Gamma$:

$$d\Gamma = \mathbf{V} \cdot \mathbf{dl} = |\mathbf{V}|\,|d\mathbf{l}|\cos\theta$$

where, θ is the angle between the vectors $\mathbf{V}$ and $\mathbf{dl}$.

The circulation around a closed curve C is the line integral:

$$\Gamma = \oint_C \mathbf{V} \cdot d\mathbf{l}$$

The dimensions of circulation are length squared, divided by time; $L^2 \cdot T^{-1}$, which is equivalent to velocity times length.

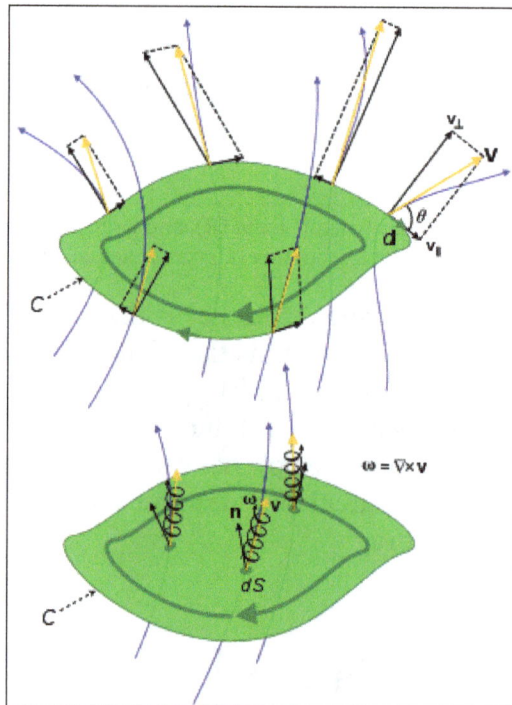

Field lines of a vector field v, around the boundary of an open curved surface with infinitesimal line element dl along boundary, and through its interior with dS the infinitesimal surface element and n the unit normal to the surface. Top: Circulation is the line integral of v around a closed loop C. Project v along dl, then sum. Here v is split into components perpendicular & parallel to dl, the parallel components are tangential to the closed loop and contribute to circulation, the perpendicular components do not. Bottom: Circulation is also the flux of vorticity ω through the surface, and the curl of v is *heuristically* depicted as a helical arrow (not a literal representation). Note the projection of v along dl and curl of v may be in the negative sense, reducing the circulation.

Kutta–Joukowski Theorem

The lift per unit span (L') acting on a body in a two-dimensional inviscid flow field can be expressed as the product of the circulation Γ about the body, the fluid density ρ, and the speed of the body relative to the free-stream **V**. Thus,

$$L' = \rho V \Gamma$$

This is known as the Kutta–Joukowski theorem.

This equation applies around airfoils, where the circulation is generated by airfoil action; and around spinning objects experiencing the Magnus effect where the circulation is induced mechanically. In airfoil action, the magnitude of the circulation is determined by the Kutta condition.

Circulation is often used in computational fluid dynamics as an intermediate variable to calculate forces on an airfoil or other body. When an airfoil is generating lift the circulation around the airfoil is finite, and is related to the vorticity of the boundary layer. Outside the boundary layer the vorticity is zero everywhere and therefore the circulation is the same around every circuit, regardless of the length of the circumference of the circuit.

Relation to Vorticity

Circulation can be related to vorticity:

$$\omega = \nabla \times \mathbf{V}$$

by Stokes' theorem:

$$\Gamma = \oint_{\partial S} \mathbf{V} \cdot d\mathbf{l} = \int\int_S \omega \cdot d\mathbf{S}$$

where the closed integration path (indicated by "∂S") is the boundary or perimeter of a surface S whose local perpendicular unit vector is $d\mathbf{S}$. Thus vorticity is the circulation per unit area, taken around a local infinitesimal loop. Correspondingly, the flux of vorticity vectors through a surface S is equal to the circulation around its perimeter.

Momentum

Momentum is a derived quantity, calculated by multiplying the mass, m (a scalar quantity), times velocity, v (a vector quantity). This means that the momentum has a direction and that direction is always the same direction as the velocity of an object's motion. The variable used to represent momentum is p. The equation to calculate momentum is shown below:

Equation for Momentum

$$p = mv$$

The SI units of momentum are kilograms times meters per second, or kg .m/s.

Vector Components and Momentum

As a vector quantity, momentum can be broken down into component vectors. When you are looking at a situation on a three-dimensional coordinate grid with directions labeled x, y, and z. For example, you can talk about the component of momentum that goes in each of these three directions:

$$p_x = mv_x$$
$$p_y = mv_y$$
$$p_z = mv_z$$

These component vectors can then be reconstituted together using the techniques of vector mathematics, which includes a basic understanding of trigonometry. Without going into the trig specifics, the basic vector equations are shown below:

$$p = p_x + p_y + p_z = mv_x + mv_y + mv_z$$

A very important application of Momentum in aerodynamics and hydrodynamics is the calculation of the drag of a body by calculating the momentum loss in its wake profile.

Airfoil in Free Flight

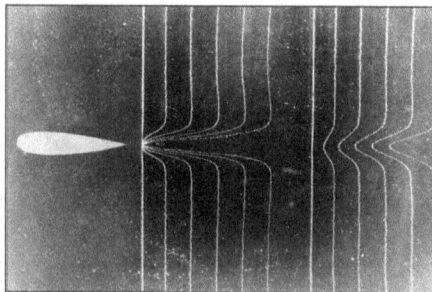

Wake profiles of an airfoil. The profiles are made visible in water flow by pulsing a voltage through a straight wire perpendicular to the flow, creating small bubbles of hydrogen that subsequently move downstream with the flow.

When fluid moves past a body the viscosity of the fluid causes it to stick to the surface of the body (no-slip condition). This creates very large velocity gradients normal to the surface of the body. These velocity gradients give rise to viscous shear stresses according to Newton's Law of viscosity:

$$\tau = \mu \frac{du}{dy}$$

These shear stresses, integrated over the surface of the body result in skin friction drag. The sum of skin friction drag and pressure (form) drag, which results from flow separation on the body, is known as the profile drag of the body.

From Newton's 3rd Law, the body then exerts a force of equal magnitude but in opposite direction on the fluid, reducing its momentum. Hence, when a fluid moves past a body we observe that the momentum of the fluid directly behind the body is significantly reduced, as shown in figure.

Top and Bottom Control Surfaces Chosen to be Streamsurfaces

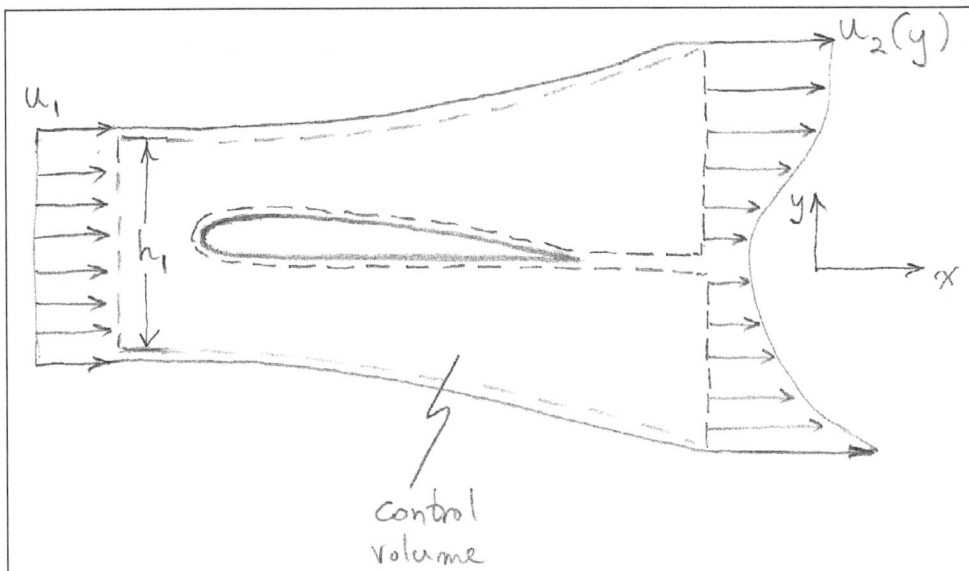

Schematic of the control volume around an airfoil in free flight. The airfoil is excluded from the control volume.

Assumptions:

- Steady flow,

- Incompressible flow,

- 2- D flow,

- Viscous effects along the control surface are negligible.

What else is going on:

- One flow stream,

- Non-uniform velocity profile downstream,

- Flow speed changes,

- Flow cross-sectional area changes.

The selection of the control volume is key in simplifying the calculation.

- The airfoil is excluded from the control volume, so that the forces *FH and FV* from the airfoil on the air will show up as external forces on the control volume. From Newton's 3rd Law:

$$\vec{F}_H = -\vec{D}$$
$$\vec{F}_V = -\vec{L}$$

where $\vec{D}$ and $\vec{L}$ are the drag and lift forces of the airfoil. If the airfoil were included in the control volume, these four forces would all be internal to the control volume and cancel each other out. As a result, none of these forces would appear in Momentum.

- Streamsurfaces are used for the upper and lower bounds of the control volume, so there is no flow crossing the top and bottom control surface.

- The control surface is taken far enough from the body in all directions, so that the pressure on it may be assumed uniform, equal to ambient atmospheric pressure:

$$p = p_\infty$$

- The width of the control volume into the paper may be taken as b, the wingspan or simply as one unit of length.

Continuity

$$\dot{m}_1 = \dot{m}_2$$
$$\dot{m}_1 = \rho u_1 A_1 = \rho u_1 \left(h_1 b\right)$$

$A_2 \succ A_1$ to maintain the same mass flow rate because the fluid slows down behind the airfoil. Also, since the velocity profile at 2 is non-uniform, an integral is needed to describe the mass flow rate there:

$$\dot{m}_2 = \rho b \int_2 u_2 \left(y\right) dy$$

Momentum

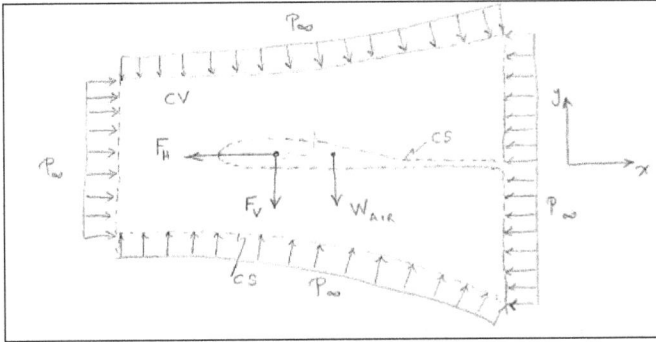

Freebody diagram for the control volume surrounding the airfoil:
p indicates the local pressure along the control surface, while the
arrows around the control surface indicate local pressure forces.

The airfoil exerts a viscous force (through its boundary layer) on the fluid flowing past
and it slows it down, so:

$$\dot{M}_{x,out} \prec \dot{M}_{x,in}$$

x – Momentum: $F_H = \dot{M}_{x,out} - \dot{M}_{x,in}$

The velocity profile at 1 is uniform, so we can write the momentum there using an algebraic expression:

$$\dot{M}_{x,in} = \rho u_1^2 \left(bh_1 \right)$$

The velocity profile at 2 is non-uniform and we need an integral to describe the momentum there:

$$\dot{M}_{x,out} = \rho b \int_2 u_2^2 \left(y \right) dy$$

x – Momentum: $-F_H = D = \rho b \int_2 u_2^2 \left(y \right) dy - \rho b h_1 u_1^2$

Top and Bottom Control Surfaces Chosen Parallel to the x - axis

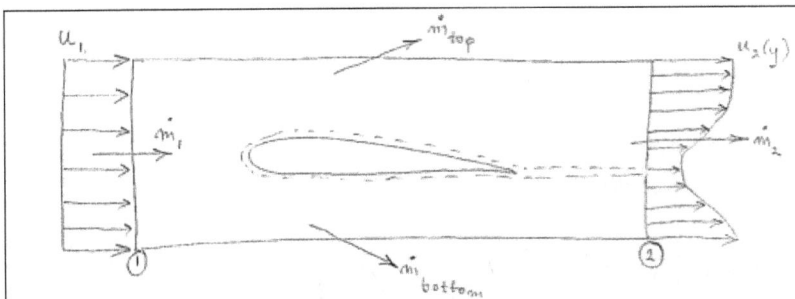

Schematic of a different control volume around an airfoil in
free flight. The airfoil is again excluded from the control volume.

Assumptions:

- Steady flow,

- Incompressible flow,

- 2- D flow,

- Viscous effects along the control surface are negligible.

What else is going on:

- One flow stream,

- Non-uniform velocity profile downstream,

- Flow speed changes,

- Constant flow cross-sectional area: $A_1 = A_2$.

This control volume appears simpler in shape than the one chosen before. However, it is now more complex to write Continuity and Momentum. The top and bottom control surface are not streamsurfaces. This implies that there is mass and momentum leakage from both of these surfaces, so: $\dot{m}_2 \prec \dot{m}_1$

Continuity

$A_2 = A_1 \rightarrow$ for the flow to be steady, i.e. no mass accumulation inside the control volume:

$$\dot{m}_1 = \dot{m}_{top} + \dot{m}_{bottom} + \dot{m}_2$$

as before $\dot{m}_1 = \rho u_1 A_1 = \rho u_1 (h_1 b)$

and

$$\dot{m}_2 = \rho b \int_2 u_2 (y) dy$$

$$\dot{m}_{top} + \dot{m}_{bottom} = \rho u_1 (h_1 b) - \rho b \int_2 u_2 (y) dy$$

So,

$$\dot{m}_{top} + \dot{m}_{bottom} = \frac{1}{2} \left[\rho u_1 (h_1 b) - \rho b \int_2 u_2 (y) dy \right]$$

Momentum

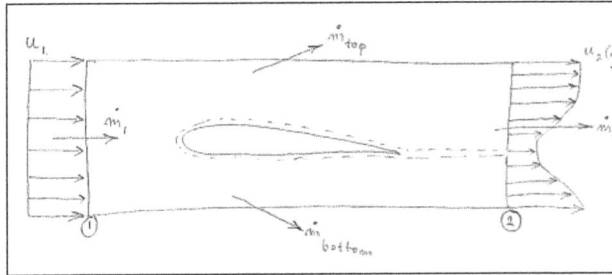

Free body diagram for the control volume surrounding the airfoil: p_∞ indicates the local pressure along the control surface, while the arrows around the control surface indicate local pressure forces.

x = Momentum: $F_H = \dot{M}_{x,out} - \dot{M}_{x,in} = \dot{M}_{x,out,2} + \dot{M}_{x,out,top+bottom} - \dot{M}_{x,in}$

$$\dot{M}_{x,in} = \rho u_1^2 \left(b h_1 \right)$$

$$\dot{M}_{x,out} = \rho b \int_2 u_2^2 \left(y \right) dy$$

It is assumed that the flow velocity through the top and bottom control surface is only slightly inclined with respect to the x – axis, hence:

$$u_{top} = u_{bottom} \simeq u_1$$

$$\dot{M}_{x,out,top+bottom} \simeq 2 \dot{m}_{top} u_1$$

x – Momentum: $-F_H = D = \rho b \int_2 u_2^2 \left(y \right) dy + 2 \dot{m}_{top} u_1 - \rho b h_1 u_1^2$

where $\dot{m}_{top}$ top was calculated above from Continuity.

Non-uniform Pressure in the Wake

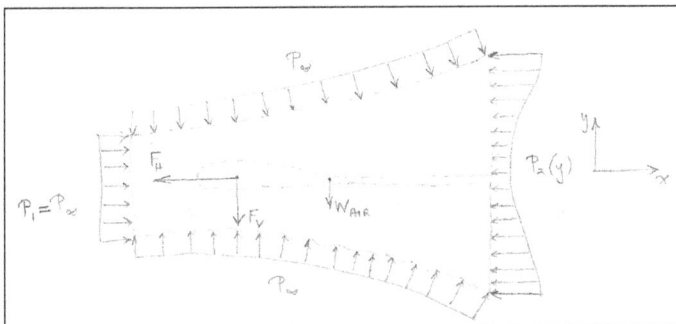

Freebody diagram for the control volume surrounding the airfoil. p_∞ and $p_2 \left(y \right)$ indicate the local pressure along the control surface, while the arrows around the control surface indicate local pressure forces.

If the control surface at the exit of the control volume is close to the airfoil the pressure at 2 will be affected by the presence of the airfoil and $p_1 \neq p_2$. Now in addition to the viscous force from the airfoil there is also a pressure force on the control volume.

x – Momentum:

$$F_H + \left[p_1 A_1 + p_1 \left(A_2 - A_1 \right) - b \int_2 p_2 \left(y \right) dy \right] = \rho b \int_2 u_2^2 \left(y \right) dy - \rho b h_1 u_2^2$$

It has been assumed that the pressure on the top and bottom portions of the control surface has remained equal to atmospheric pressure. The pressure forces on each of these surfaces (top + bottom) act normal to the surface. If we take the x – component of these pressure forces at each point of the top and bottom surfaces we get $p_1 \left(A_2 - A_1 \right)$, which is precisely the 2nd term in the bracket in the x - Momentum. In this case the drag is calculated from:

$$-F_H = D = \rho b \int_2 u_2^2 \left(y \right) dy - \rho b h_1 u_1^2 + b \left[p_1 h_2 - \int_2 p_2 \left(y \right) dy \right]$$

Airfoil in a Wind Tunnel Test Section

The easiest way to measure airfoil drag is in a wind tunnel. To avoid 3-D (tip) effects airfoils are mounted from wall to wall in the test section of a wind tunnel.

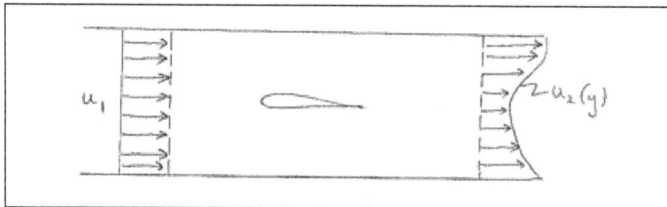

Schematic of an airfoil in a wind tunnel test section.

Assumptions:

- Steady flow,
- Incompressible flow,
- 2 – D flow,
- Viscous effects along the wind tunnel walls are neglected.

What else is going on:

- One flow stream,
- Non-uniform velocity profile downstream,

- Flow speed changes from streamline to streamline,

- Flow cross-sectional area remains constant: $A_1 = A_2$.

Since the airfoil is inside a wind tunnel the top and bottom control surface are by default the wind tunnel walls. The flow cannot cross these walls, hence they must be streamsurfaces of the flow. In a real wind tunnel flow, the no slip condition requires zero velocity at each wall. This condition, however, does not contradict having the top and bottom walls as streamsurfaces of the flow.

Continuity

$$\dot{m}_{in} = \dot{m}_{out}$$

$$\rho u_1 A = \rho \int_2 u_2(y)\, dy = \rho \bar{u}_2 A$$

Hence, no matter what the wake profile looks like, Continuity requires that the average velocity at 2 is the same as the average velocity at 1.

NB: $u_1 = \bar{u}_2$ and $\bar{u}_2 = \dfrac{1}{h}\displaystyle\int_2 u_2(y)\, dy$ and $\overline{u_2^2} = \dfrac{1}{h}\displaystyle\int_2 u_2^2(y)\, dy$

So, $\bar{u}_2^2 \neq \overline{u_2^2} \Rightarrow \dot{M}_{x,1} = \rho u_1^2 A \neq \rho \overline{u_2^2} A = \dot{M}_{x,2}$

Momentum

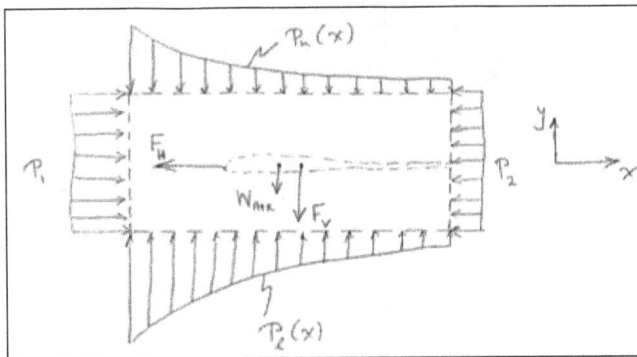

Freebody diagram for the control volume surrounding an airfoil in a wind tunnel test section: $p_1, p_u(x), p_l(x), p_2$ indicate local pressure along the control surface, while the arrows around the control surface indicate local pressure forces.

x – Momentum: $-F_H = D = \rho b \displaystyle\int_2 u_2^2(y)\, dy - \rho b h u_1^2 + (p_2 - p_1) b h$

The pressure in the downstream section has been assumed constant across the height of the test section but not the same as the pressure upstream. Space constraints in the wind tunnel dictate measuring the wake traverse fairly close to the airfoil and this results in $p_2 \neq p_1$.

The pressure on the top and bottom wind tunnel walls (as well as along the top and bottom control surface) cannot be the same if the airfoil is lifting. When an airfoil lifts it pushes the air down, hence, we expect the pressure on the lower wall to be higher than the pressure on the upper wall. These pressure forces must be included in the y – Momentum equation:

$$y - \text{Momentum: } F_V + b\left(\int\limits_{Lower\ CS} p_l(x)dx - \int\limits_{Upper\ CS} P_u(x)dx \right) = \dot{M}_{y,out} - \dot{M}_{y,in} = 0$$

Obviously, there is no vertical momentum crossing the control surface at any point, so our control surface is in equilibrium in the y – direction, as shown by the y – Momentum. Hence the lift of the airfoil is:

$$-F_V = L = b\left(\int\limits_{Lower\ CS} p_l(x)dx - \int\limits_{Upper\ CS} P_u(x)dx \right)$$

Can be found by integrating and subtracting the p – distributions on the bottom and top wind tunnel walls.

Momentum Effects on Aerodynamic Forces

Lift is created by deflecting a flow of air, and drag is generated on a body in a wide variety of ways. From Newton's second law of motion, the aerodynamic force F on the body is directly related to the change in momentum of the fluid with time t. The fluid momentum is equal to the mass m times the velocity V of the fluid.

$$F = d\,(m \times V)/dt$$

$$F = \text{constant} \times V \times m/t$$

Since the air moves, defining the mass is tricky. If the mass of fluid were brought to a halt, it would occupy some volume in space. We can define the density (r) of the fluid to be the mass divided by the volume v.

$r = m/v$

Since the fluid is moving, we must determine the mass in terms of the mass flow rate. The mass flow rate is the amount of mass passing a given point during some time interval t and its units are mass/time. We can relate the mass flow rate to the density mathematically. The mass flow rate mdot is equal to the density times the velocity times the area A through which the mass passes.

$mdot = m/t = r \times V \times A$

With knowledge of the mass flow rate, we can express the aerodynamic force as equal to the mass flow rate times the velocity.

$F = constant \times V \times r \times V \times A$

A quick units check:

$mass \times length/time^2 = constant \times length/time \times mass/length^3 \times length/time \times length^2$

$mass \times length/time^2 = mass \times length/time^2$

Combining the velocity dependence and absorbing the area into the constant, we find:

$F = constant \times r \times V^2$

The aerodynamic force equals a constant times the density times the velocity squared. The dynamic pressure of a moving flow is equal to one half of the density times the velocity squared. The aerodynamic force is directly proportional to the dynamic pressure of the flow.

Effect of Velocity on Aerodynamic Forces

The velocity used in the aerodynamic equation is the relative velocity between an object and the flow. The aerodynamic force depends on the square of the velocity. Doubling the velocity quadruples the force. The dependence of lift and drag on the square of the velocity has been known for more than a hundred years. The Wright brothers used this information in the design of their first aircraft.

Effect of Air Density on Aerodynamic Forces

The aerodynamic force depends linearly on the density of the air. Halving the density halves the force. As altitude increases, the air density decreases. This explains why airplanes have a flight ceiling, an altitude above which it cannot fly. As an airplane ascends, a point is reached where there is not enough air mass to generate enough lift to overcome the airplane's weight. The relation between altitude and density is a fairly complex exponential.

Conservation of Energy

$w = \text{Work} = p_2 v_2 - p_1 v_1 + w_{sh}$

$h = \text{Enthalpy} = e + pv$ $e = \text{Internal Energy}$

$q = \text{Heat Flow}$ $u = \text{velocity}$

$k = \text{Kinetic Energy} = \dfrac{u^2}{2}$ $v = \text{volume}$

$p = \text{pressure}$ $w_{sh} = \text{shaft work}$

1st Law of Thermodynamics:

$$e_2 - e_1 + k_2 - k_1 = q - w_{sh} - p_2 v_2 + p_1 v_1$$

(algebra):

$$e_2 + p_2 v_2 - e_1 - p_1 v_1 + (u^2/2)_2 - (u^2/2)_1 = q - w_{sh}$$

(more algebra):

$$h_2 + (u^2/2)_2 - h_1 - (u^2/2)_1 = q - w_{sh}$$

define:

$$h + (u^2/2) = h_t = \text{Total Enthalpy}$$

$$\boxed{h_{t_2} - h_{t_1} = q - w_{sh}}$$

The conservation of energy is a fundamental concept of physics along with the conservation of mass and the conservation of momentum. Within some problem domain, the amount of energy remains constant and energy is neither created nor destroyed. Energy can be converted from one form to another (potential energy can be converted to kinetic energy) but the total energy within the domain remains fixed.

Thermodynamics is a branch of physics which deals with the energy and work of a system. As mentioned on the gas properties slide, thermodynamics deals only with the large scale response of a system which we can observe and measure in experiments.

On some separate slides, we have discussed the state of a static gas, the properties which define the state, and the first law of thermodynamics as applied to any system, in general. On this slide we derive a useful form of the energy conservation equation for a gas beginning with the first law of thermodynamics. If we call the internal energy of a gas E, the work done by the gas W, and the heat transferred into the gas Q, then the first law of thermodynamics indicates that between state "1" and state "2":

$$E_2 - E_1 = Q - W$$

Aerospace engineers usually simplify a thermodynamic analysis by using intensive variables; variables that do not depend on the mass of the gas. We call these variables specific variables. We create a "specific" variable by taking a property whose value depends on the mass of the system and dividing it by the mass of the system. Many of the state properties listed on this slide, such as the work and internal energy depend on the total mass of gas. We will use "specific" versions of these variables. Engineers usually use the lower case letter for the "specific" version of a variable. Our first law equation then becomes:

$$e_2 - e_1 = q - w$$

Because we are considering a moving gas, we add the specific kinetic energy term to the internal energy on the left side. The normal kinetic energy K of a moving substance is equal to 1/2 times the mass m times the velocity u squared:

$$K = (m \times u^2)/2$$

Then the specific kinetic energy k is given by:

$$k = (u^2)/2$$

and the first law equation becomes:

$$e_2 - e_1 + k_2 - k_1 = q - w$$

There are two parts to the specific work for a moving gas. Some of the work, called the shaft work (wsh) is used to move the fluid or turn a shaft, while the rest of the work goes into changing the state of the gas. For a pressure p and specific volume v, the work is given by:

$$w = (p \times v)2 - (p \times v)1 + wsh$$

Substituting:

$$e_2 - e_1 + k_2 - k_1 = q - (p \times v)2 + (p \times v)1 - wsh$$

If we perform a little algebra on the first law of thermodynamics, we can begin to group some terms of the equations:

$$e_2 + (p \times v)2 - e_1 - (p \times v)1 + [(u^2)/2]2 - [(u^2)/2]1 = q - wsh$$

A useful additional state variable for a gas is the specific enthalpy h which is equal to:

$$h = e + (p \times v)$$

Simplifying the energy equation:

$$h_2 - h_1 + [(u^2)/2]2 - [(u^2)/2]1 = q - wsh$$

or

$$h_2 + [(u^2)/2]2 - h_1 - [(u^2)/2]1 = q - wsh$$

By combining the velocity terms with the enthalpy terms to form the total specific enthalpy "h_t" we can further simplify the equation.

$$h_t = h + u^2/2$$

The total specific enthalpy is analogous to the total pressure in Bernoulli's equation; both expressions involve a "static" value plus one half the square of the velocity.

The final, most useful, form of the energy equation is given in the red box.

$$h_{t_2} - h_{t_1} = q - w_{sh}$$

For a compressor or power turbine, there is no external heat flow into the gas and the "q" term is set equal to zero. In the burner, no work is performed and the "wsh" term is set to zero.

Center of Pressure

The Centre of Pressure is the average location of all of the pressure acting upon a body moving through a fluid.

As an aircraft moves through the atmosphere, the velocity of the air varies around the surfaces of the aircraft. As an example, the air is accelerated as it passes over the cambered surfaces of the aerofoils. This variation of air velocity, especially over the wing and tail surfaces, produces a variation in the local pressure at various places on the aircraft. The average location of the pressure variation is referred to as the centre of pressure. The total aerodynamic force can be considered to act through the centre of pressure and can be resolved into its two components, lift and drag.

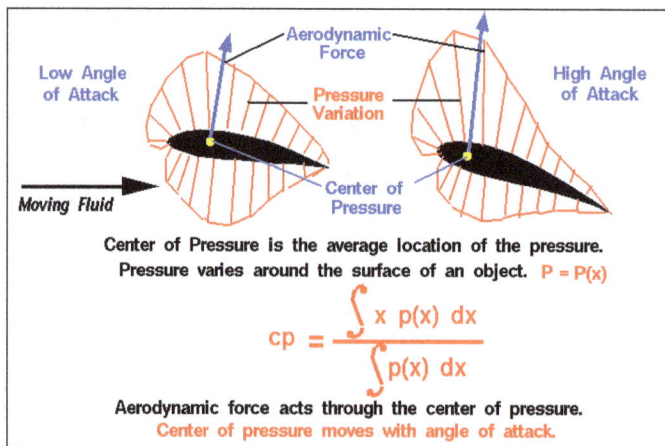

Center of Pressure is the average location of the pressure.
Pressure varies around the surface of an object. $P = P(x)$

$$cp = \frac{\int x\, p(x)\, dx}{\int p(x)\, dx}$$

Aerodynamic force acts through the center of pressure.
Center of pressure moves with angle of attack.

The centre of pressure does not remain in a constant location. As the angle of attack changes, the local pressure at every point on the aerofoil also changes. This, in turn, causes a change in the location of the center of pressure.

Determining the center of pressure is very important for any flying object. To trim an airplane, or to provide stability for a model rocket or a kite, it is necessary to know the location of the center of pressure of the entire aircraft.

In general, determining the center of pressure (cp) is a very complicated procedure because the pressure changes around the object. Determining the center of pressure requires the use of calculus and a knowledge of the pressure distribution around the body. We can characterize the pressure variation around the surface as a function p(x) which indicates that the pressure depends on the distance x from a reference line usually taken as the leading edge of the object. If we can determine the form of the function, there are methods to perform a calculus integration of the equation. We will use the symbols "S[ ]dx" to denote the integration of a continuous function. Then the center of pressure can be determined from:

$$cp = (S[x \times p(x)]dx) / (S[p(x)]dx)$$

If we don't know the actual functional form, we can numerically integrate the equation using a spreadsheet by dividing the distance into a number of small distance segments and determining the average value of the pressure over that small segment. Taking the sum of the average value times the distance times the distance segment divided by the sum of the average value times the distance segment will produce the center of pressure.

There are several important problems to consider when determining the center of pressure for an airfoil. As we change angle of attack, the pressure at every point on the airfoil changes. And, therefore, the location of the center of pressure changes as well. The movement of the center of pressure caused a major problem for early airfoil designers because the amount (and sometimes the direction) of the movement was different for different designs. In general, the pressure variation around the airfoil also imparts a torque, or "twisting force", to the airfoil. If a flying airfoil is not restrained in some way it will flip as it moves through the air.

To resolve some of these design problems, aeronautical engineers prefer to characterize the forces on an airfoil by the aerodynamic force, coupled with an aerodynamic moment to account for the torque. It was found both experimentally and analytically that, if the aerodynamic force is applied at a location 1/4 chord back from the leading edge on most low speed airfoils, the magnitude of the aerodynamic moment remains nearly constant with angle of attack. Engineers call the location where the aerodynamic moment remains constant the aerodynamic center of the airfoil. Using the aerodynamic center as the location where the aerodynamic force is applied eliminates the problem of the movement of the center of pressure with angle of attack in aerodynamic analysis. (For supersonic airfoils, the aerodynamic center is nearer the 1/2 chord location.)

When computing the trim of an aircraft, model rocket, or kite, we usually apply the aerodynamic forces at the aerodynamic center of airfoils and compute the center of pressure of the vehicle as an area-weighted average of the centers of the components.

Incompressible Flow

Incompressible flow, in general terms of fluid mechanics, refers to a fluid that maintains constant density during a flow. To an extent, all fluid flows have some change in density when subjected to an external force or internal viscous forces; however, density variation is more prevalent in some analysis results than others. The ability to keep density constant in equations of fluid dynamics can greatly reduce the computational burden, and by assuming a flow as incompressible, the engineer is stating that this technique will not significantly compromise the accuracy of the solution. The incompressible flow assumption has large implications within aerospace engineering, and the application in which the analysis is being performed will be essential in determining whether the assumption can be utilized.

Most incompressible flows within aerospace engineering are in the field of aerodynamics, where compressibility effects of air flow can be neglected if the Mach number is below 0.3. In this velocity range, the maximum change in density of air is less than 5%, so it is assumed that this variation is negligible. Beyond Mach 0.3, the variation in density can no longer be ignored, and analysis must involve the principles of compressible flow. Liquids, in most cases, can be regarded as incompressible fluids because of the immense force required to change their density. There are, however, certain applications of liquid flows in which compressibility effects cannot be ignored, such as water hammer, cavitation, and high-pressure hydraulics. The underlying principle of the incompressible flow assumption is that all materials (solids, liquids, gases) are compressible, even if the change in density is on the order of 10^{-9}. In some instances, it is very clear that density can remain constant in an analysis without creating inaccuracies in the solution. In other instances, it becomes a judgment call of the engineer, and that is why incompressible flow is considered an assumption rather than an actual physical occurrence.

Incompressible and Inviscid Flow

Friction within a flow is introduced by the fluid's viscosity, or the extent to which its particles interact to hinder flow. Inviscid flow is an assumption in which these frictional effects internal to the flow are neglected. Another representation of this assumption is with the Reynolds number, which approaches infinity for inviscid flow (the inertial effects greatly outweigh the viscous effects). Incompressible, inviscid flow is one of the fundamental foundations in fluid mechanics. Simply put, it is the most simplified type of flow because it does not introduce compressible or friction effects. This becomes noticeably helpful in the above equations, where density can be taken outside the differential in the left-side terms and the friction forces disappear on the right side. Simplification of these terms also diminishes the need to calculate Reynolds, number, surface relative roughness, etc. When flow is considered steady, or the properties don't change with time, the Navier-Stokes equations can be reduced to:

$$\rho \frac{D\vec{V}}{Dt} = \rho\vec{f} - \nabla p$$

This equation is known as Euler's equation, and relates changes in pressure to changes in velocity. This is the basis dynamic pressure, q, which is involved in many aerodynamics equations:

$$q = \frac{1}{2}\rho V_2$$

Euler's equation can also be transformed to Bernoulli's equation, which relates flow properties at two points on the same streamline.

Flow Along a Streamline

Integrating Euler's equation along a streamline results in:

$$\frac{p}{\rho} + \frac{V^2}{2} + gz = \text{constant}$$

Which is known as Bernoulli's equation, and can easily be transformed into an energy form? This equation is very powerful because changes in pressure, velocity, and elevation are interrelated and can be solved for at any point on the streamline. Relating two points on a streamline is as easy as:

$$\frac{p_1}{\rho} + \frac{V_1^2}{2} + gz_1 = \frac{p_2}{\rho} + \frac{V_2^2}{2} + gz_2$$

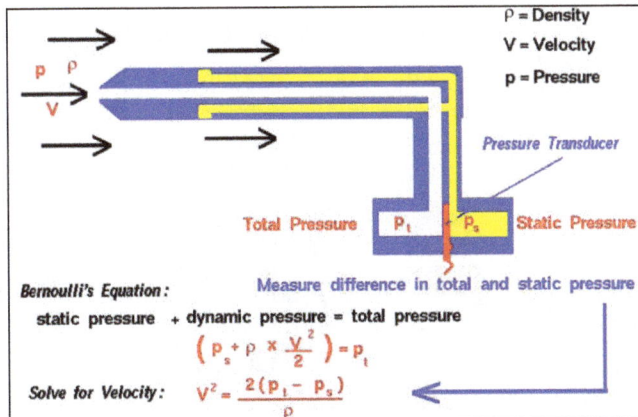

Bernoulli's equation is the principle in being able to measure velocity with a pitot tube. This is because it directly relates pressure to velocity.

This form can be used to analyze piping systems at two different points, which is the basis for one of the flow measurement techniques mentioned below (pitot tube). It is very important, however, to understand when these equations apply.

Validity of Incompressible Flows

There is a large difference in validity of incompressible flow that depends on whether the flow is internal or external. Internal flows are more likely to be exposed to large external forces, which may introduce compressibility effects in both liquids and gases (pneumatics, water hammer, hydraulics, gas pipelines, etc.). For external flows, especially air flow across a wing section, and other flows relevant to aerospace engineering, incompressible flow is valid for low speeds (M < 0.3). In respect to the previous equations, validity is achieved only under the following restrictions:

- Steady flow,

- Incompressible flow,

- Inviscid flow,

- Flow along a streamline (+).

(+) This condition is only needed for Bernoulli's equation.

As an important reminder, incompressible and inviscid flows are assumptions and do not physically occur in nature. This principle is more important at a microscopic level rather than macroscopic, such as large fixed wing aircraft design. Slight density changes in microfluidics could have a greater impact than they would at a larger scale. Hence, it is very important to know the application before assuming a flow compressible or incompressible.

Aerospace Engineering Applications

Many types of flows within aerospace engineering can benefit from an incompressible flow, constant density analysis.

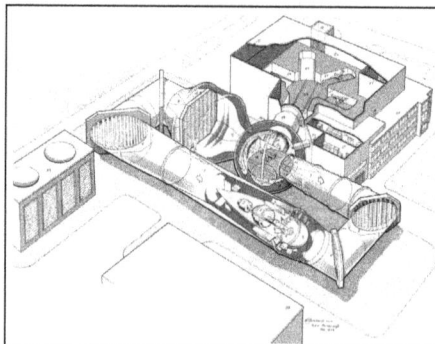

Correct wind tunnel speeds are achieved by utilizing Bernoulli's equation as well.

Low-speed Aerodynamics

The most prevalent application is within low-speed aerodynamics. The mechanics of flight are easily calculated when the incompressible flow is assumed. These flows include:

- Atmospheric flow over an airfoil (wing) during flight, especially NLF airfoils which operate best at low speeds.

- Testing components/models in a low-speed wind tunnel.

- Analyzing flow through low-speed wind tunnel convergences/divergences.

- Velocity measurement through pitot tubes.

- Flow over most wind turbine blades.

Aircraft Components

Liquid flows through machinery play a large role in aircraft performance and its abilities. Many of these applications are modeled as incompressible flows, which include:

- Fuel flow from reservoir to aircraft engine,

- Flow through hydraulic lines (with small loads).

Since the scope of an aerospace engineer does not lie only in the skies, other applications such as water flow about Naval ship hulls can be assumed to be incompressible.

Compressible Flow

When a fluid flow is compressible, the fluid density varies with its pressure. Compressible flows are usually high speed flows with Mach numbers greater than about 0.3. Examples include aerodynamic applications such as flow over a wing or aircraft nacelle as well as industrial applications such as flow through high-performance valves.

Incompressible flows do not have such a variation of density. The key differentiation between compressible and incompressible is the velocity of the flow. A fluid such as air that is moving slower than Mach 0.3 is considered incompressible, even though it is a gas. A gas that is run through a compressor is not truly considered compressible (in the thermodynamic sense) unless its velocity exceeds Mach 0.3. This is important to note because analyses run as compressible can be harder to run, and require longer analysis times than incompressible flows.

Subsonic compressible flows have a Mach number between 0.3 and 0.8. The relationship between pressure and density is weak, and no shocks will be computed within the flow.

Compressible flows have a Mach number greater than 0.8. The pressure strongly affects the density, and shocks are possible. Compressible flows can be either transonic (0.8 < M < 1.2) or supersonic (1.2 < M < 3.0). In supersonic flows, pressure effects are only transported downstream. The upstream flow is not affected by conditions and obstructions downstream.

The speed of sound is given as a:

$$a = \sqrt{\gamma R T}$$

where,

$\gamma = 1.4$ for air, R = gas constant, and T = reference static temperature (in absolute units).

The velocity, V, is then the product of the sound speed, a, and the Mach number, M:

$$V = aM$$

The total temperature, T_t, is a key parameter as well, and is the sum of the static temperature and the dynamic temperature. There are two way to calculate total temperature:

$$T_t = T + \frac{V_i^2}{2C_p} \text{ or } T_t = T\left(1 + \frac{\gamma-1}{2}M^2\right)$$

- V is the velocity.

- C_p is the gas specific heat.

For air, $C_p = 1005 \text{ m}^2/(\text{s}^2 \text{ K})$.

The total temperature must be specified as a constant value for analyses that do not have heat transfer and as a boundary condition for those that do.

The total pressure, Pt, is another useful quantity for running compressible analyses. It is the sum of the static pressure and the dynamic pressure.

If the flow accelerates through a geometrically converging section to sonic speed, the flow is considered to be choked. When choked, no additional mass can pass through the constriction region, even as the pressure drop is increased (by lowering the outlet back pressure). The flow downstream of the throat can then expand and become supersonic.

Basic Solution Strategy

Compressible flow analyses are much more sensitive to the applied boundary conditions and material properties than incompressible analyses. If the applied settings do not define a physically real flow situation, then the analysis can be very unstable and may fail to reach a converged solution.

For this reason, we recommend that you understand the flow situation that you are trying to analyze. Proper specification of the boundary conditions and material properties will greatly improve the chances of a successful analysis.

Test-runs

A technique that can be very helpful when starting a new analysis is to mock up a two dimensional representation of the model to ensure that all conditions are correct. Inconsistent settings will be revealed very quickly as a 2D model is run, allowing for much faster debugging of the analysis. When the settings properly define the analysis, they can be applied to the (usually) much bigger 3D (or more detailed 2D) model with confidence that any additional adjustments to the model have to be made to the mesh and not the fundamental set-up.

Meshing

To capture physical elements such as shocks, the mesh size will have to be quite fine in critical areas. The mesh can be less fine in non-critical areas. A good guideline governing mesh transition is that the mesh size should not transition by more than a factor of four between neighboring fluid volumes. In general, a coarse mesh will be more stable but less accurate. For this reason, as part of the test procedure described above, in some cases it is recommended to verify the analysis set-up with a coarse mesh, and then when you are confident in your settings, refine the mesh to improve accuracy.

Materials

To allow the density to vary, open the Material Environment dialog, and select Variable. If the operating conditions are different from the default values, right click on the Material branch of the Design Study bar, and click Edit environment reference. Specify the appropriate static pressure and static temperature. Because the density is calculated using these values, the Environment pressure needs to be exact for the gage reference point to be correct.

Heat Transfer

To include heat transfer in a compressible analysis, apply Total (stagnation) temperature boundary conditions instead of static temperatures at the inlets. Total temperature should also be applied to any solids or walls with known temperature conditions. (Do not use a Static Temperature boundary condition to define a known temperature in a compressible analysis. At a wall the value of static and total temperature is the same, and should be applied as a total temperature.) The Set Heat Transfer to on in the Solve dialog. The value of Total Temperature on the Solve dialog will be ignored if heat transfer is enabled.

When heat transfer is present in a compressible analysis, viscous dissipation, pressure work, and kinetic energy terms are calculated. It is only necessary to enable heat transfer if you are solving for heat transfer or for flow velocities higher than M = 3.0 if viscous dissipation is important or to capture a very crisp shock.

It is very important that the total temperature is specified correctly. A good test is to run zero iterations and check that the Mach number at the inlet is the expected value. If not, adjust the total temperature and inlet boundary conditions accordingly.

If heat transfer is not solved for, it is necessary to specify a Total temperature in the Solve dialog. The equation for total temperature is given above.

Absolute, Total, Static and Dynamic Values

The term absolute is used in conjunction with pressure. Normally, the solution to the pressure equation is a relative pressure. This relative pressure does not contain the gravitational head or the rotational head or the reference pressure. It is the part of the pressure that is affected by the velocities in the momentum equation directly. The absolute pressure adds the gravitational and rotational heads and the reference pressure to that calculated from the pressure equation. Referring to the relative pressure as Prel, the absolute pressure is calculated as:

$$P_{absolute} = P_{rel} + P_{ref} + \rho_{ref} \sum_i g_i X_i + \rho_{ref} \sum_i \omega_i^2 X_i^2$$

where, the *ref* subscript refers to reference values, the subscript *i* refers to the 3 coordinate directions, *g* is the gravitational acceleration and ω is the rotational speed. The reference density is calculated at the beginning of the analysis using the reference pressure and temperature. For flows with a constant density, the reference density is the constant value. For flows which have no gravitational or rotational heads, the relative pressure is the gage pressure.

The terms dynamic and static are used most commonly with compressible fluids. The dynamic values are kinetic energy-like terms:

$$T_{dynamic} = \frac{V^2}{2c_p}$$

$$P_{dynamic} = \frac{1}{2} \rho V^2$$

The specific heat used to calculate the dynamic temperature is not the thermal value entered on the property window, but is a mechanical value calculated using:

$$c_p = \frac{\gamma R_{gas}}{\gamma - 1}$$

Where γ is the ratio of the constant pressure specific heat to the constant volume specific heat and R_{gas} is the gas constant for this gas.

The static temperature is determined by solving the energy equation. For adiabatic properties, the energy equation that is used to determine the static temperature is the constant total temperature equation. Hence, the static temperature is the total or stagnation temperature minus the dynamic temperature.

The static pressure is the absolute pressure The total temperature is the sum of the static and dynamic temperatures. The total pressure is the sum of the static or absolute pressure and the dynamic pressure.

Transonic Flow

When an airplane is in motion at subsonic speeds, the air is treated as though it was incompressible. As airplane speed increases, however, the air loses its assumed incompressibility and the error in estimating, for example, drag, becomes greater and greater.

The question arises as to how fast an airplane must be moving before one must take into account compressibility.

A disturbance in the air will send pressure pulses or waves out into the air at the speed of sound. Consider the instance of a cannon fired at sea level. An observer situated some distance from the cannon will see the flash almost instantaneously, but the sound wave is heard (or the pressure wave is felt) sometime later. The observer can easily compute the speed of sound by dividing the distance between him and the cannon by the time it takes the sound to reach him. The disturbance propagates out away from the cannon in an expanding hemispherical shell.

The speed of sound varies with altitude. More precisely, it depends upon the square root of the absolute temperature. At sea level under standard conditions (T_0 = 288.15 K (degrees Kelvin)), the speed of sound is 340.3 meters per second (761.2 miles per hour), but at an altitude of 15 kilometers (9.3 miles or 49,212 feet) where the temperature is down to 216.7 K, the speed of sound is only 295.1 meters per second (660.2 miles per hour). This difference indicates that an airplane flying at this altitude encounters the speed of sound at a slower speed, and, therefore, comes up against compressibility effects sooner.

An airplane flying well below the speed of sound creates a disturbance in the air and sends out pressure pulses in all directions. Air ahead of the airplane receives these "messages" before the airplane arrives and the flow separates around the airplane. But as the plane approaches the speed of sound, the pressure pulses merge closer and closer together in front of the airplane and little time elapses between the time the air gets a warning of the plane's approach and the plane's actual arrival time at the speed of sound, the pressure pulses move at the same speed as the plane. They merge ahead of the airplane into a "shock wave" that is an almost instantaneous line of change in pressure, temperature, and density. The air has no warning of the approach of the airplane and abruptly passes through the shock system. There is a tendency for the air to break away from the airplane and not flow smoothly about it; as a result, there is a change in the aerodynamic forces from those experienced at low incompressible flow speeds.

The Mach number is a measure of the ratio of the airplane speed to the speed of sound. In other words, it is a number that may relate the degree of warning that air may have to an airplane's approach. The Mach number is named after Ernst Mach, an Austrian professor. For Mach numbers less than one, one has subsonic flow, for Mach numbers greater than one, supersonic flow, and for Mach numbers greater than 5, the name is hypersonic flow. Additionally, transonic flow pertains to the range of speeds in which flow patterns change from subsonic to supersonic or vice versa, about Mach 0.8 to 1.2. Transonic flow presents a special problem area as neither equations describing subsonic flow nor those describing supersonic flow may be accurately applied to the regime.

At subsonic speeds, drag was composed of three main components—skin-friction drag, pressure drag, and induced drag (or drag due to lift). At transonic and supersonic speeds, there is a substantial increase in the total drag of the airplane due to funda-mental changes in the pressure distribution.

This drag increase encountered at these high speeds is called *wave drag*. The drag of the airplane wing, or for that matter, any part of the airplane rises sharply, and large increases in thrust are necessary to obtain further increases in speed. This wave drag is due to the unstable formation of shock waves that transforms a considerable part of the available propulsive energy into heat, and to the induced separation of the flow from the airplane surfaces. Throughout the transonic range, the drag coefficient of the airplane is greater than in the supersonic range because of the erratic shock formation and general flow instabilities. Once a supersonic flow has been established, however, the flow stabilizes and the drag coefficient is reduced.

The total drag at transonic and supersonic speeds can be divided into two categories: (1) zero-lift drag composed of skin-friction drag and wave (or pressure-related) drag of zero lift and (2) lift-dependent drag composed of induced drag (drag due to lift) and wave (or pressure-related) drag due to lift. In the early days of transonic flight, the sound barrier represented a real barrier to higher speeds. Once past the transonic re-gime, the drag coefficient and the drag decrease, and less thrust is required to fly super-sonically. However, as it proceeds toward higher supersonic speeds, the drag increases (even though the drag coefficient may show a decrease).

It is a large loss in propulsive energy due to the formation of shocks that causes wave drag. Up to a free-stream Mach number of about 0.7 to 0.8, compressibility effects have only minor effects on the flow pattern and drag. The flow is subsonic everywhere. As the flow must speed up as it proceeds about the airfoil, the local Mach number at the airfoil surface will be higher than the free-stream Mach number. There eventually occurs a free-stream Mach number called the critical Mach number at which a supersonic point appears somewhere on the airfoil surface, usually near the point of maximum thickness, and indicates that the flow at that point has reached Mach 1. As the free-stream Mach number is increased beyond the critical Mach number and approaches Mach 1, larger and larger regions of supersonic flow appear on the airfoil surface. In order for this supersonic flow to return to subsonic flow, it must pass through a shock (pressure discontinuity). This loss of velocity is accompanied by an increase in temperature, that is, a production of heat. This heat represents an expenditure of propulsive energy that may be presented as wave drag. These shocks appear anywhere on the airplane (wing, fuselage, engine nacelles, etc.) where, due to curvature and thickness, the localized Mach number exceeds 1.0 and the airflow must decelerate below the speed of sound. For transonic flow, the wave drag increase is greater than would be estimated from a loss of energy through the shock. In fact, the shock wave interacts with the boundary layer so that a separation of the boundary layer occurs immediately behind the shock. This condition accounts for a large increase in drag that is known as shock-induced (boundary-layer) separation.

The free-stream Mach number at which the drag of the airplane increases markedly is called the *drag-divergence Mach number*. Large increases in thrust are required to produce any further increases in airplane speed. If an airplane has an engine of insufficient thrust, its speed will be limited by the drag-divergence Mach number. The prototype Convair F-102A was originally designed as a supersonic interceptor but early flight tests indicated that because of high drag, it would never achieve this goal. It later achieved its goal through a redesign.

At a free-stream Mach number greater than 1, a bow shock appears around the airfoil nose. Most of the airfoil is in supersonic flow. The flow begins to realign itself parallel to the body surface and stabilize, and the shock-induced separation is reduced.

This condition results in lower drag coefficients. Supersonic flow is better behaved than transonic flow and there are adequate theories that can predict the aerodynamic forces and moments present. Often, in transonic flow, the flow is unsteady, and the shock waves on the body surface may jump back and forth along the surface, thus disrupting and separating the flow over the wing surface. This sends pulsing, unsteady flow back to the tail surfaces of the airplane. The result is that the pilot feels a buffeting and vibration of both wing and tail controls. This condition occurred especially in the first airplane types to probe the sound barrier. With proper design, however, airplane configurations gradually evolved to the point where flying through the transonic region posed little or no difficulty in terms of wing buffeting or loss of lift.

The question arises as to whether one may delay the drag-divergence Mach number to a value closer to 1 so as to impart the ability to fly at near-sonic velocities with the same available engine thrust before encountering large wave drag. There are a number of ways of delaying the transonic wave drag rise (or equivalently, increasing the drag-divergence Mach number closer to 1). These include:

- Use of thin airfoils;

- Use of a forward or backward swept wing;

- Low-aspect-ratio wing;

- Removal of boundary layer and vortex generators; and

- Supercritical and area-rule technology.

Supersonic Flow

Many of the techniques used to delay transonic drag rise also are directly applicable in designing the airplane to fly with minimum wave drag in the supersonic regime.

A bow shock wave will exist for free-stream Mach numbers above 1.0. In three dimensions, the bow shock is in reality a cone in shape (a Mach cone) as it extends back from the nose of the airplane. As long as the wing is swept back behind the Mach cone, there is subsonic flow over most of the wing and relatively low drag. A delta wing has the advantage of a large sweep angle but also greater wing area than a simple swept wing to compensate for the loss of lift usually experienced in sweepback. But, at still higher supersonic Mach numbers, the Mach cone may approach the leading edge of even a highly swept delta wing. This condition causes the total drag to increase rapidly and, in fact, a straight wing (no sweep) becomes preferable.

Sweepback has been used primarily in the interest of minimizing transonic and supersonic wave drag. At subsonic Mach numbers, however, the disadvantages dominate. They include high induced drag (due to small wing span or low aspect ratio), high angles of attack for maximum lift, and reduced effectiveness of trailing-edge flaps. The straight-wing airplane does not have these disadvantages. For an airplane that is designed to be multimission, for example, cruise at both subsonic and supersonic velocities, it would be advantageous to combine a straight wing and swept wing design. This is the logic for the variable sweep or swing-wing. Although not necessarily equal to the optimum configurations in their respective speed regimes, it is evident that an airplane with a swing-wing capability can, in a multimissioned role, over the total speed regime, be better than the other airplanes individually. One major drawback of the swing-wing airplane is the added weight and complexity of the sweep mechanisms. But technological advances are solving these problems also.

In addition to low-aspect-ratio wings at supersonic speeds, supersonic wave drag may also be minimized by using thin wings and area ruling. Also, long, slender, cambered fuselages minimize drag and improve the spanwise lift distribution.

Choked Flow

Choked flow is a compressible flow effect. The parameter that becomes "choked" or "limited" is the fluid velocity.

Choked flow is a fluid dynamic condition associated with the Venturi effect. When a flowing fluid at a given pressure and temperature passes through a constriction (such as the throat of a convergent-divergent nozzle or a valve in a pipe) into a lower pressure environment the fluid velocity increases. At initially subsonic upstream conditions, the conservation of mass principle requires the fluid velocity to increase as it flows through the smaller cross-sectional area of the constriction. At the same time, the Venturi effect causes the static pressure, and therefore the density, to decrease at the constriction. Choked flow is a limiting condition where the mass flow will not increase with a further decrease in the downstream pressure environment for a fixed upstream pressure and temperature.

For homogeneous fluids, the physical point at which the choking occurs for adiabatic conditions, is when the exit plane velocity is at sonic conditions; i.e., at a Mach number of 1. At choked flow, the mass flow rate can be increased only by increasing density upstream and at the choke point.

The choked flow of gases is useful in many engineering applications because the mass flow rate is independent of the downstream pressure, and depends only on the temperature and pressure and hence the density of the gas on the upstream side of the restriction. Under choked conditions, valves and calibrated orifice plates can be used to produce a desired mass flow rate.

Choked Flow in Liquids

If the fluid is a liquid, a different type of limiting condition (also known as choked flow) occurs when the Venturi effect acting on the liquid flow through the restriction causes a decrease of the liquid pressure beyond the restriction to below that of the liquid's vapor pressure at the prevailing liquid temperature. At that point, the liquid will partially flash into bubbles of vapor and the subsequent collapse of the bubbles causes cavitation. Cavitation is quite noisy and can be sufficiently violent to physically damage valves, pipes and associated equipment. In effect, the vapor bubble formation in the restriction prevents the flow from increasing any further.

Mass Flow Rate of a Gas at Choked Conditions

All gases flow from upstream higher pressure sources to downstream lower pressure sources. There are several situations in which choked flow occurs, such as the change of the cross section in a de Laval nozzle or the flow through an orifice plate. Here the most important part is where to calculate the choked velocity: at upstream or downstream of a nozzle or orifice. The choked velocity is always observed at upstream of an orifice or nozzle and this velocity is usually less than the speed of sound in air. Another important aspect is that this is the actual velocity of the upstream fluid. Hence, the upstream actual volumetric flow rate, when expanded to downstream pressure, will result in a more actual volumetric flow for the downstream condition. Thus, the overall leakage rate when measured at downstream conditions needs to take care of this fact. When this choked velocity has reached the mass flow rate from upstream to downstream, it can still be increased if the upstream pressure is increased. However, this value of the choked velocity will keep the actual volumetric flow rate (Actual Gas Flow rate, and hence velocity) the same irrespective of the upstream pressure, provided that choked flow conditions prevail.

Choking in Change of Cross Section Flow

Assuming ideal gas behavior, steady-state choked flow occurs when the downstream pressure falls below a critical value p^*. That critical value can be calculated from the dimensionless critical pressure ratio equation,

$$\frac{p^*}{p_0} = \left(\frac{2}{\gamma+1} \right)^{\frac{\gamma}{\gamma-1}}$$

where γ is the heat capacity ratio c_p / c_v of the gas and where p_0 is the total (stagnation) upstream pressure.

For air with a heat capacity ratio $\gamma = 1.4$, then $p^* = 0.528 p_0$ other gases have γ in the range 1.09 (e.g. butane) to 1.67 (monatomic gases), so the critical pressure ratio varies in the range $0.487 < p^* / p_0 < 0.587$, which means that, depending on the gas, choked flow usually occurs when the downstream static pressure drops to below 0.487 to 0.587 times the absolute pressure in stagnant upstream source vessel.

When the gas velocity is choked, the equation for the mass flow rate is:

$$\dot{m} = C_d A \sqrt{\gamma \rho_0 P_0 \left(\frac{2}{\gamma+1} \right)^{\frac{\gamma+1}{\gamma-1}}}$$

where:

$\dot{m}$, mass flow rate, in kg/s,

C_d, discharge coefficient, dimensionless,

A, discharge hole cross-sectional area, in m²,

$\gamma, \dfrac{c_p}{c_v}$ of the gas,

c_p, specific heat of the gas at constant pressure,

c_v, specific heat of the gas at constant volume,

ρ_0, real gas (total) density at total pressure P_0 and total temperature T_0, in kg/m³,

P_0, absolute upstream total pressure of the gas, in Pa, or kg/m·s²,

T_0, absolute upstream total temperature of the gas, in K.

The mass flow rate is primarily dependent on the cross-sectional area A of the nozzle throat and the upstream pressure P, and only weakly dependent on the temperature T. The rate does not depend on the downstream pressure at all. All other terms are constants that depend only on the composition of the material in the flow. *Although the gas velocity reaches a maximum and becomes choked, the mass flow rate is not choked.* The mass flow rate can still be increased if the upstream pressure is increased as this increases the density of the gas entering the orifice.

The value of C_d can be calculated using the below expression:

$$C_d = \frac{\dot{m}}{A\sqrt{2\rho\Delta P}}$$

where,

C_d, discharge coefficient through the constriction (dimensionless),

A, cross-sectional area of flow constriction (unit length squared),

$\dot{m}$, mass flow rate of fluid through constriction (unit mass of fluid per unit time),

ρ, density of fluid (unit mass per unit volume),

ΔP, pressure drop across constriction (unit force per unit area).

The above equations calculate the steady state mass flow rate for the pressure and temperature existing in the upstream pressure source.

If the gas is being released from a closed high-pressure vessel, the above steady state equations may be used to approximate the *initial* mass flow rate. Subsequently, the mass flow rate will decrease during the discharge as the source vessel empties and the pressure in the vessel decreases. Calculating the flow rate versus time since the initiation of the discharge is much more complicated, but more accurate. Two equivalent methods for performing such calculations are explained and compared online.

The technical literature can be very confusing because many authors fail to explain whether they are using the universal gas law constant R which applies to any ideal gas or whether they are using the gas law constant R_s which only applies to a specific individual gas. The relationship between the two constants is $R_s = R / M$ where M is the molecular weight of the gas.

Real Gas Effects

If the upstream conditions are such that the gas cannot be treated as ideal, there is no closed form equation for evaluating the choked mass flow. Instead, the gas expansion should be calculated by reference to real gas property tables, where the expansion takes place at constant enthalpy.

Minimum Pressure Ratio Required for Choked Flow to Occur

The minimum pressure ratios required for choked conditions to occur (when some typical industrial gases are flowing) are presented in table. The ratios were obtained using the criterion that choked flow occurs when the ratio of the absolute upstream pressure to the absolute downstream pressure is equal to or greater than $(2 / [\gamma + 1])^{-\gamma/(\gamma-1)}$, where γ is the specific heat ratio of the gas. The minimum pressure ratio may be understood as the ratio between the upstream pressure and the pressure at the nozzle throat when the gas is traveling at Mach 1; if the upstream pressure is too low compared to the downstream pressure, sonic flow cannot occur at the throat.

Gas	$\gamma = \dfrac{c_p}{c_v}$	Min. P_u/P_d for choked flow
Dry air	1.400 at 20 °C	1.893
Nitrogen	1.404 at 15 °C	1.895
Oxygen	1.400 at 20 °C	1.893
Helium	1.660 at 20 °C	2.049
Hydrogen	1.410 at 20 °C	1.899
Methane	1.307	1.837
Propane	1.131	1.729
Butane	1.096	1.708
Ammonia	1.310 at 15 °C	1.838
Chlorine	1.355	1.866
Sulfur dioxide	1.290 at 15 °C	1.826
Carbon monoxide	1.404	1.895
Carbon dioxide	1.30	1.83

- P_u, absolute upstream gas pressure.

- P_d, absolute downstream gas pressure.

Venturi Nozzles with Pressure Recovery

The flow through a venturi nozzle achieves a much lower nozzle pressure than downstream pressure. Therefore, the pressure ratio is the comparison between the upstream and nozzle pressure. Therefore, flow through a venturi can reach Mach 1 with a much lower upstream to downstream ratio.

Thin-plate Orifices

The flow of real gases through thin-plate orifices never becomes fully choked. The mass flow rate through the orifice continues to increase as the downstream pressure is lowered to a perfect vacuum, though the mass flow rate increases slowly as the downstream pressure is reduced below the critical pressure. Cunningham first drew attention to the fact that choked flow will not occur across a standard, thin, square-edged orifice.

Vacuum Conditions

In the case of upstream air pressure at atmospheric pressure and vacuum conditions downstream of an orifice, both the air velocity and the mass flow rate becomes choked or limited when sonic velocity is reached through the orifice.

The Flow Pattern

Figure shows the flow through the nozzle when it is completely subsonic (i.e. the nozzle is not choked). The flow in the chamber accelerates as it converges toward the throat, where it reaches its maximum (subsonic) speed at the throat. The flow then decelerates through the diverging section and exhausts into the ambient as a subsonic jet. Lowering the back pressure, in this state, will increase the flow speed everywhere in the nozzle.

When the back pressure, p_b, is lowered enough, the flow speed is Mach 1 at the throat, as in figure. The flow pattern is exactly the same as in subsonic flow, except that the flow speed at the throat has just reached Mach 1. Flow through the nozzle is now choked since further reductions in the back pressure can't move the point of M=1 away from the throat. However, the flow pattern in the diverging section does change as you lower the back pressure further.

As p_b is lowered below that needed to just choke the flow, a region of supersonic flow forms just downstream of the throat. Unlike in subsonic flow, the supersonic flow accelerates as it moves away from the throat. This region of supersonic acceleration is terminated by a normal shock wave. The shock wave produces a near-instantaneous deceleration of the flow to subsonic speed. This subsonic flow then decelerates through the remainder of the diverging section and exhausts as a subsonic jet. In this regime if you lower or raise the back pressure you move the shock wave away from (increase the length of supersonic flow in the diverging section before the shock wave) the throat.

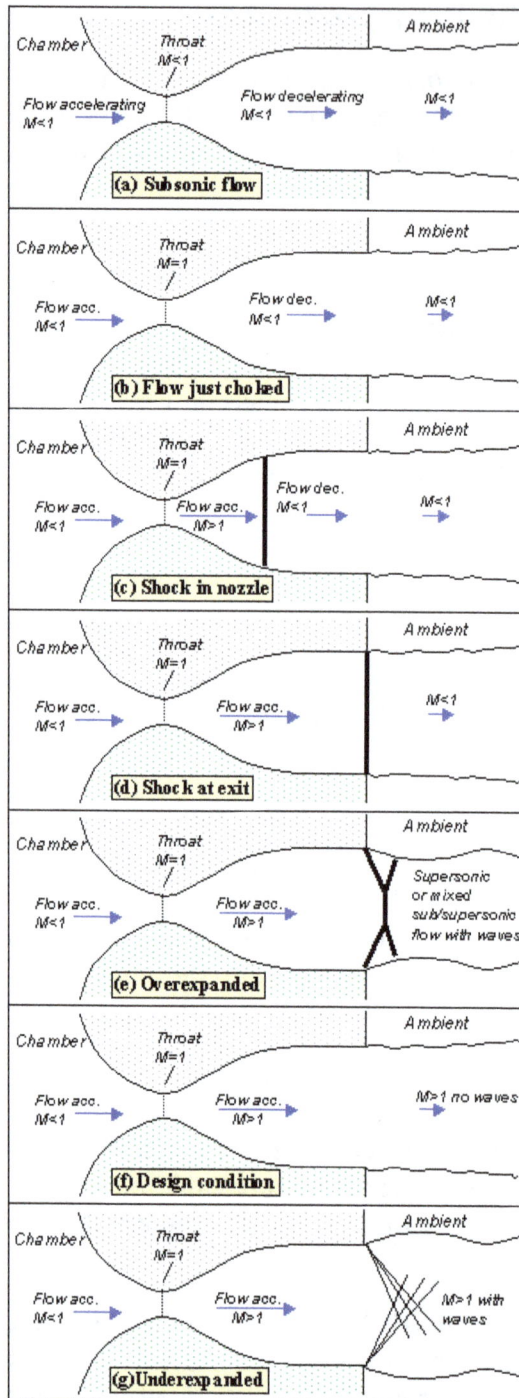

(a) Subsonic flow

(b) Flow just choked

(c) Shock in nozzle

(d) Shock at exit

(e) Overexpanded

(f) Design condition

(g) Underexpanded

If the p_b is lowered enough, the shock wave will sit at the nozzle exit. Due to the very long region of acceleration (the entire nozzle length) the flow speed will reach its maximum just before the shock front. However, after the shock the flow in the jet will be subsonic.

Lowering the back pressure further causes the shock to bend out into the jet, and a complex pattern of shocks and reflections is set up in the jet which will involve a mixture of subsonic and supersonic flow, or (if the back pressure is low enough) just supersonic flow. Because the shock is no longer perpendicular to the flow near the nozzle walls, it deflects the flow inward as it leaves the exit producing an initially contracting jet. This is referred as overexpanded flow because in this case the pressure at the nozzle exit is lower than that in the ambient (the back pressure)- i.e. the flow has been expanded by the nozzle too much.

A further lowering of the back pressure changes and weakens the wave pattern in the jet. Eventually the back pressure will be low enough so that it is now equal to the pressure at the nozzle exit. In this case, the waves in the jet disappear altogether, and the jet will be uniformly supersonic. This situation, since it is often desirable, is referred to as the 'design condition'.

Finally, if the back pressure is lowered even further we will create a new imbalance between the exit and back pressures (exit pressure greater than back pressure). In this situation (called 'under expanded') what we call expansion waves (that produce gradual turning perpendicular to the axial flow and acceleration in the jet) form at the nozzle exit, initially turning the flow at the jet edges outward in a plume and setting up a different type of complex wave pattern.

Hypersonic Flow

In aerodynamics, a Hypersonic Flow is one that greatly exceeds the speed of sound, often stated as starting at speeds of Mach 5 and above.

The precise Mach number at which a craft can be said to be flying at Hypersonic Flow varies, since individual physical changes in the airflow (like molecular dissociation and ionization) occur at different speeds; these effects collectively become important around Mach 5-10. The hypersonic regime is often alternatively defined as speeds where C_p and C_v are no longer able to be reasonably considered constant.

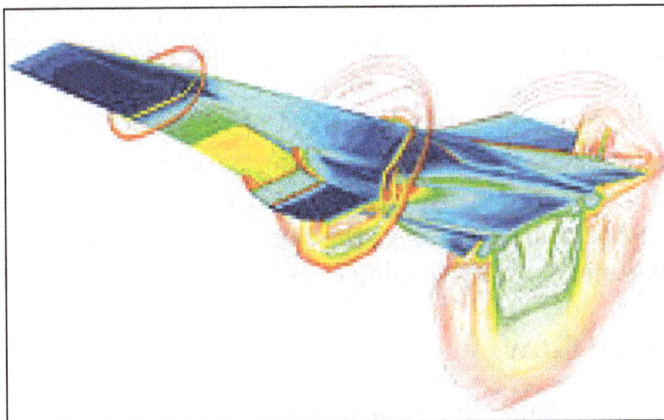

Characteristics of Flow

While the definition of hypersonic flow can be quite vague and is generally debatable (especially due to the absence of discontinuity between supersonic and hypersonic flows), a hypersonic flow may be characterized by certain physical phenomena that can no longer be analytically discounted as in supersonic flow. The peculiarity in hypersonic flows are as follows:

- Shock layer,
- Aerodynamic heating,
- Entropy layer,
- Real gas effects,
- Low density effects,
- Independence of aerodynamic coefficients with Mach number.

Small Shock Stand-off Distance

As a body's Mach number increases, the density behind a bow shock generated by the body also increases, which corresponds to a decrease in volume behind the shock due to conservation of mass. Consequently, the distance between the bow shock and the body decreases at higher Mach numbers.

Entropy Layer

As Mach numbers increase, the entropy change across the shock also increases, which results in a strong entropy gradient and highly vortical flow that mixes with the boundary layer.

Viscous Interaction

A portion of the large kinetic energy associated with flow at high Mach numbers transforms into internal energy in the fluid due to viscous effects. The increase in internal energy is realized as an increase in temperature. Since the pressure gradient normal to the flow within a boundary layer is approximately zero for low to moderate hypersonic Mach numbers, the increase of temperature through the boundary layer coincides with a decrease in density. This causes the bottom of the boundary layer to expand, so that the boundary layer over the body grows thicker and can often merge with the shock wave near the body leading edge.

High-temperature Flow

High temperatures due to a manifestation of viscous dissipation cause non-equilibrium chemical flow properties such as vibrational excitation and dissociation and ionization of molecules resulting in convective and radiative heat-flux.

Classification of Mach Regimes

Although "subsonic" and "supersonic" usually refer to speeds below and above the local speed of sound respectively, aerodynamicists often use these terms to refer to particular ranges of Mach values. This occurs because a "transonic regime" exists around M=1 where approximations of the Navier–Stokes equations used for subsonic design no longer apply, partly because the flow locally exceeds M=1 even when the freestream Mach number is below this value.

The "supersonic regime" usually refers to the set of Mach numbers for which linearised theory may be used; for example, where the (air) flow is not chemically reacting and where heat transfer between air and vehicle may be reasonably neglected in calculations.

Generally, NASA defines "high" hypersonic as any Mach number from 10 to 25, and re-entry speeds as anything greater than Mach 25. Among the aircraft operating in this regime are the Space Shuttle and (theoretically) various developing space-planes.

In the following table, the "regimes" or "ranges of Mach values" are referenced instead of the usual meanings of "subsonic" and "supersonic".

| Regime | Velocity | | | | General plane characteristics |
	Mach No	mph	km/h	m/s	
Subsonic	< 0.8	< 614	< 988	< 274	Most often propeller-driven and commercial turbofan aircraft with high aspect-ratio (slender) wings, and rounded features like the nose and leading edges.
Transonic	0.8–1.2	614–921	988–1482	274–412	Transonic aircraft nearly always have swept wings that delay drag-divergence, and often feature designs adhering to the principles of the Whitcomb area rule.
Supersonic	1.2–5	921–3836	1482–6174	412–1715	Aircraft designed to fly at supersonic speeds show large differences in their aerodynamic design because of the radical differences in the behavior of fluid flows above Mach 1. Sharp edges, thin airfoil-sections, and all-moving tailplane/canards are common. Modern combat aircraft must compromise in order to maintain low-speed handling; "true" supersonic designs include the F-104 Starfighter and BAC/Aérospatiale Concorde.

Hypersonic	5–10	3836–7673	6174–12350	1715–3430	Cooled nickel or titanium skin; highly integrated (due to domination of interference effects: non-linear behavior means that superposition of results for separate components is invalid), small wings, see X-51A Waverider, HyperSoar and WU-14 (DF-ZF).
High-hyper-sonic	10–25	7673–19180	12350–30870	3430–8575	Thermal control becomes a dominant design consideration. Structure must either be designed to operate hot, or be protected by special silicate tiles or similar. Chemically reacting flow can also cause corrosion of the vehicle's skin, with free-atomic oxygen featuring in very high-speed flows. Examples include the 53T6 ABM-3 Gazelle (Mach 17) anti-ballistic missile, the DF-41 (Mach 25) intercontinental ballistic missile and the Russian Avangard hypersonic vehicle (Mach 27). Hypersonic designs are often forced into blunt configurations because of the aerodynamic heating rising with a reduced radius of curvature.
Re-entry speeds	> 25	> 19180	> 30870	> 8575	Ablative heat shield; small or no wings; blunt shape.

Similarity Parameters

The categorization of airflow relies on a number of similarity parameters, which allow the simplification of a nearly infinite number of test cases into groups of similarity. For transonic and compressible flow, the Mach and Reynolds numbers alone allow good categorization of many flow cases.

Hypersonic flows, however, require other similarity parameters. First, the analytic equations for the oblique shock angle become nearly independent of Mach number at high (~>10) Mach numbers. Second, the formation of strong shocks around aerodynamic bodies means that the freestream Reynolds number is less useful as an estimate of the behavior of the boundary layer over a body (although it is still important). Finally, the increased temperature of hypersonic flows means that real gas effects become important. For this reason, research in hypersonics is often referred to as aerothermodynamics, rather than aerodynamics.

The introduction of real gas effects means that more variables are required to describe the full state of a gas. Whereas a stationary gas can be described by three variables (pressure, temperature, adiabatic index), and a moving gas by four (flow velocity), a hot gas in chemical equilibrium also requires state equations for the chemical components of the gas, and a gas in nonequilibrium solves those state equations using time as an extra variable. This means that for a nonequilibrium flow, something between 10 and 100 variables may be required to describe the state of the gas at any given time. Additionally, rarefied hypersonic flows (usually defined as those with a Knudsen number above 0.1) do not follow the Navier–Stokes equations.

Hypersonic flows are typically categorized by their total energy, expressed as total enthalpy (MJ/kg), total pressure (kPa-MPa), stagnation pressure (kPa-MPa), stagnation temperature (K), or flow velocity (km/s).

Wallace D. Hayes developed a similarity parameter, similar to the Whitcomb area rule, which allowed similar configurations to be compared.

Regimes

Hypersonic flow can be approximately separated into a number of regimes. The selection of these regimes is rough, due to the blurring of the boundaries where a particular effect can be found.

Perfect Gas

In this regime, the gas can be regarded as an ideal gas. Flow in this regime is still Mach number dependent. Simulations start to depend on the use of a constant-temperature wall, rather than the adiabatic wall typically used at lower speeds. The lower border of this region is around Mach 5, where ramjets become inefficient, and the upper border around Mach 10-12.

Two-temperature Ideal Gas

This is a subset of the perfect gas regime, where the gas can be considered chemically perfect, but the rotational and vibrational temperatures of the gas must be considered separately, leading to two temperature models. See particularly the modeling of supersonic nozzles, where vibrational freezing becomes important.

Dissociated Gas

In this regime, diatomic or polyatomic gases (the gases found in most atmospheres) begin to dissociate as they come into contact with the bow shock generated by the body. Surface catalysis plays a role in the calculation of surface heating, meaning that the type of surface material also has an effect on the flow. The lower border of this regime is where any component of a gas mixture first begins to dissociate in the stagnation point of a flow (which for nitrogen is around 2000 K). At the upper border of this regime, the effects of ionization start to have an effect on the flow.

Ionized Gas

In this regime the ionized electron population of the stagnated flow becomes significant, and the electrons must be modeled separately. Often the electron temperature is handled separately from the temperature of the remaining gas components. This region occurs for freestream flow velocities around 10–12 km/s. Gases in this region are modeled as non-radiating plasmas.

Radiation-dominated Regime

Above around 12 km/s, the heat transfer to a vehicle changes from being conductively dominated to radiatively dominated. The modeling of gases in this regime is split into two classes:

- Optically thin: Where the gas does not re-absorb radiation emitted from other parts of the gas.

- Optically thick: Where the radiation must be considered a separate source of energy.

The modeling of optically thick gases is extremely difficult, since, due to the calculation of the radiation at each point, the computation load theoretically expands exponentially as the number of points considered increases.

Continuity Equation

The law of Aerodynamics are formulated by applying to a flowing gas several basic principles from physics.

Physical Principle: Mass can neither be created nor destroyed.

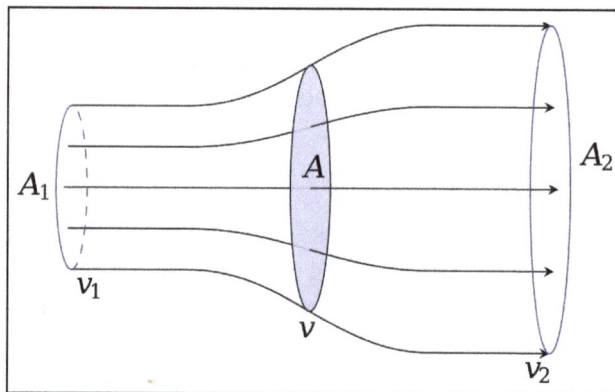

Steam tube with mass conservation.

To apply this principle of flowing gas:

- Consider an imaginary circle drawn perpendicular to the flow direction.

- Now look at all the steam lines that go through the circumference of the circle. These streamlines form a tube called Stream tube.

- As we move along with the gas confined inside the stream tube, we see that the cross sectional area of the tube may change, say in moving from point 1 to point 2.

- However, as long as the flow is steady, the mass that flows through the cross section at point 1 must be the same as the mass that flows through the cross section at point 2, become the definition of stream line, there can be no flow across streamlines.

- The mass flowing through the stream tube is confined by the streamlines of the boundary, much as the flow of water through a flexible garden hose is confined by the wall of the hose.

Let,

A_1 = Cross-sectional area of the steam tube at section 1.

Now, at a given instant in time, consider all the fluid elements that are momentarily in the plane of A_1.

After a lapse of time dt, these same fluid elements all move a distance.

Distance = V_1dt

In doing so elements have swept out a volume, downstream of point. The mass of gas dm in the volume is equal to the density times the volume that is,

$Volume = A_1 V_1 \, dt$

$dm = \rho_1 \left(A_1 V_1 dt \right)$

This is the mass of gas that has swept through the cross section at point 1 during time interval dt.

The mass flow $\dot{M}$ through area A is the mass crossing A per unit time.

Therefore, from Equation $dm = \rho_1 \left(A_1 V_1 dt \right)$ of area A_1.

Mass flow $= \left(dm / dt \right) = \dot{M}_1 = \left(\rho_1 A_1 V_1 \right)$

Also, the mass flow through the other cross section A_2, bounded by the same streamlines that go through the circumference of the A_1, is obtained in the same fashion, as

$\dot{M}_2 = \left(\rho_2 A_2 V_2 \right)$

Since, mass can neither be created nor destroyed.

We have,

$\dot{M}_1 = \dot{M}_2$

Hence,

$$\rho_1 A_1 V_1 = \rho_2 A_2 V_2$$

This is Continuity Equation for steady fluid flow.

The core value of the continuity equation is the conservation law of mass: Mass can be neither created nor destroyed.

Considering a control volume fixed in the space, the volume and the surface are constant, the fluid that flows in and out of the control volume is the only thing that will change. In order to simplify the calculation of the amount of the fluid, a crucial idea is presented: mass flow. Assuming the control volume is a rectangular box, and the flow is uniform flow, then the volume of the fluid flows into the box is the flow velocity (V_n) times the unit time step (dt) times the entrance are (A): $Volume_{fluid} = (V_n dt)A$; and the mass of that is the volume of the fluid times the density of the fluid (ρ): $Mass_{fluid} = \rho(V_n dt)A$. Therefore, we can conclude that in every time step (dt), the mass of the fluid that flows into the box is $\rho V_n A$, and this is so called the mass flow rate ($\dot{m} = \rho V_n A$.) Let's take a further step, if we divide the mass flow rate by the entrance area (A), then what we have is the mass of the fluid that flows into the box in every time step, every unit area, which is the mass flux Mass flux $= \dfrac{\dot{m}}{A} = \rho V_n$.

Now, we know how to describe the amount of the fluid in a certain region in a flow field. The next step is to outline exactly how much of the fluid that flows in, out and remains in the control volume. The understanding of the fluid amount in the control volume plays an important role in the calculation of lift, drag and even propulsion force. According to the conservation law of mass, the amount of the fluid remains in the control volume will exactly equal to the amount that flows out minus that flows in the control volume. In other words, **net mass flows out of the control volume through the control surface = time rate of decrease of mass inside the control volume.**

Here, turn the bolded words in the mathematical language:

$$\oiint_S \rho V \cdot dS = -\frac{\partial}{\partial t} \iiint_V \rho dV .$$

The LHS is the integration of the whole control surface, representing the mass of the fluid that flows in the control surface through every unit area with the velocity V (note that the velocity V is always perpendicular to the unit area dS); the RHS is the integration of the whole control volume, presenting the mass in every unit volume (ρdV), but with a minus sign in the front of the differential term with respect to time due to the definition of "rate of decrease". Then, we obtain the following formula.

$$\oiint_S \rho V \cdot dS + \frac{\partial}{\partial t} \iiint_V \rho dV = 0$$

With the application of the divergence theorem $\iint_S A dS = \iiint_V \nabla \cdot A dV$ the above equation can be rewritten in $\iiint_V \nabla \cdot (\rho V) dV + \frac{\partial}{\partial t} \iiint_V \rho dV = 0$. Combine the integration sign, then it becomes, $\iiint_V [\partial \rho \partial t + \nabla \cdot (\rho V)] dV = 0$. There is no negative mass (no consideration of something like dark mass), so the only way that can make this equation be true is the inside of the integration is zero. And this is the differential form of the continuity equation:

$$\frac{\partial \rho}{\partial t} + \nabla \cdot (\rho V) = 0$$

Overall, the only assumption that we've made for the continuity equation is that the fluid is a continuum, without considering a shock wave, the behavior of the mother nature is very close to this scenario. Therefore, the continuity equation can be applied in many circumstances, including three-dimensional flow, steady or unsteady flow, any fluid (water, air, glue, etc.), viscid or inviscid flow, and compressible or incompressible flow. It is worth to mention that, under the consideration of a steady flow, the time differential term will go to zero because the density is constant respect to time. The formula can be simplified as $\nabla \cdot (\rho V) = 0$, and this is the foundation of the famous Bernoulli's equation.

Momentum Equation

The momentum equation is a statement of Newton's Second Law and relates the sum of the forces acting on an element of fluid to its acceleration or rate of change of momentum. You will probably recognise the equation F = ma which is used in the analysis of solid mechanics to relate applied force to acceleration. In fluid mechanics it is not clear what mass of moving fluid we should use so we use a different form of the equation.

Newton's 2nd Law can be written:

> The Rate of change of momentum of a body is equal to the resultant force acting on the body, and takes place in the direction of the force.

To determine the rate of change of momentum for a fluid we will consider a streamtube as we did for the Bernoulli equation, we start by assuming that we have steady flow which is non-uniform flowing in a stream tube.

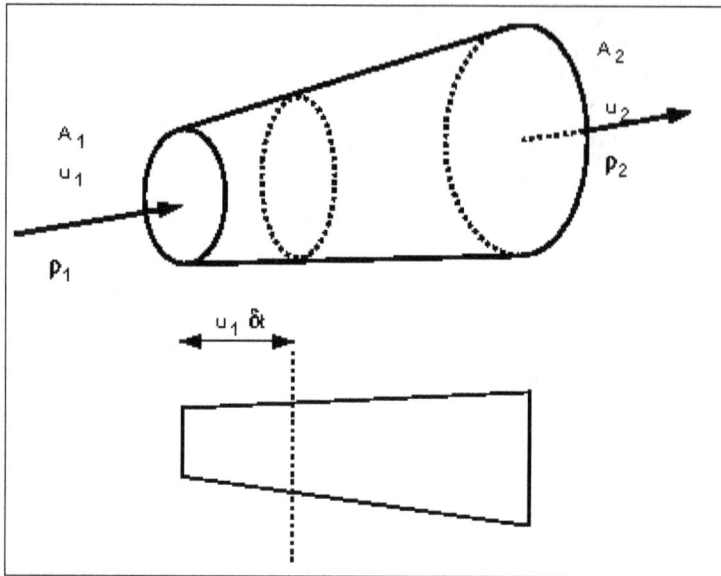

A streamtube in three and two-dimensions.

In time δt a volume of the fluid moves from the inlet a distance $u\ t$, so the volume entering the streamtube in the time δt is:

volume entering the stream tube = area × distance = $A_1 u_1 \delta t$

this has mass:

mass entering stream tube = volume × density = $\rho_1 A_1 u_1 \delta t$

and momentum:

momentum of fluid entering stream tube = mass × velocity = $\rho_1 A_1 u_1 \delta t\, u_1$

Similarly, at the exit, we can obtain an expression for the momentum leaving the steamtube:

momentum of fluid leaving stream tube = $\rho_2 A_2 u_2 \delta t\, u_2$

We can now calculate the force exerted by the fluid using Newton's 2nd Law. The force is equal to the rate of change of momentum. So,

Force = rate of change of momentum

$$F = \frac{\left(\rho_2 A_2 u_2 \delta t\, u_2 - \rho_1 A_1 u_1 \delta t\, u_1 \right)}{\delta t}$$

We know from continuity that $Q = A_1 u_1 = A_2 u_2$, and if we have a fluid of constant density, i.e. $\rho_1 = \rho_2 = \rho$, then we can write:

$$F = Q\rho(u_2 - u_1)$$

For an alternative derivation of the same expression, as we know from conservation of mass in a stream tube that:

Mass into face 1 = mass out of face 2

We can write,

$$\text{rate of change of mass} = \dot{m} = \frac{dm}{dt} = \rho_1 A_1 u_1 = \rho_2 A_2 u_2$$

The rate at which momentum leaves face 1 is:

$$\rho_2 A_2 u_2 u_2 = \dot{m} u_2$$

The rate at which momentum enters face 2 is:

$$\rho_1 A_1 u_1 u_1 = \dot{m} u_1$$

Thus the rate at which momentum changes across the stream tube is:

$$\rho_2 A_2 u_2 u_2 - \rho_1 A_1 u_1 u_1 = \dot{m} u_2 \ \dot{m} u_1$$

i.e.

Force = rate of change of momentum

$$F = \dot{m}(u_2 - u_1)$$
$$F = Q\rho(u_2 - u_1)$$

This force is acting in the direction of the flow of the fluid.

This analysis assumed that the inlet and outlet velocities were in the same direction - i.e. a one dimensional system.

Consider the two dimensional system in the figure below:

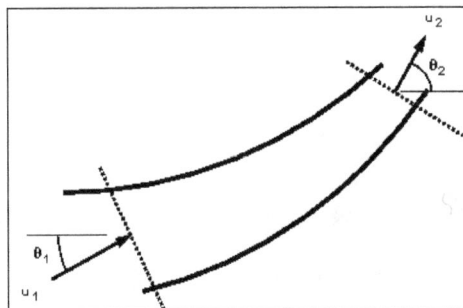

Two dimensional flow in a streamtube.

At the inlet the velocity vector, u_1, makes an angle, θ_1, with the x-axis, while at the outlet u_2 make an angle θ_2. In this case we consider the forces by resolving in the directions of the co-ordinate axes.

The force in the x-direction:

F_x = Rate of change of momentum in x – direction

= Rate of change of mass × change in velocity in x – direction

$$= \dot{m}\left(u_2\cos\theta_2 - u_1\cos\theta_1\right)$$

$$= \dot{m}\left(u_{2_x} - u_{1_x}\right)$$

$$= \rho Q\left(u_2\cos\theta_2 - u_1\cos\theta_1\right)$$

$$= \rho Q\left(u_{2_x} - u_{1_x}\right)$$

And the force in the y-direction:

$$F_y = \dot{m}\left(u_2\sin\theta_2 - u_2\sin\theta_1\right)$$

$$= \dot{m}\left(u_{2_y} - u_{1_y}\right)$$

$$= \rho Q\left(u_2\sin\theta_2 - u_1\sin\theta_1\right)$$

$$= \rho Q\left(u_{2_y} - u_{1_y}\right)$$

We then find the resultant force by combining these vectorially:

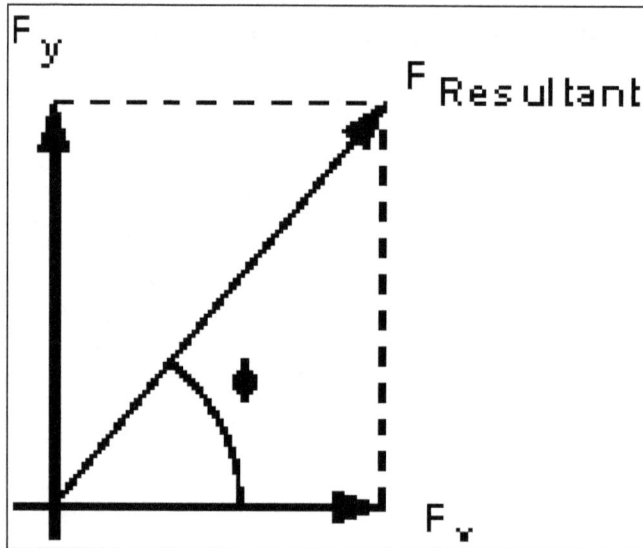

$$F_{\text{resultant}} = \sqrt{F_x^2 + F_y^2}$$

And the angle which this force acts at is given by:

$$\phi = \tan^{-1}\left(\frac{F_y}{F_x}\right)$$

For a three-dimensional (x, y, z) system we then have an extra force to calculate and resolve in the z-direction. This is considered in exactly the same way.

We can say:

The total force the fluid = rate of change of momentum through the control volume

$$F = \dot{m}\left(u_{out} - u_{in}\right)$$
$$= \rho Q\left(u_{out} - u_{in}\right)$$

Remember that we are working with vectors so F is in the direction of the velocity. This force is made up of three components:

F_R = Force exerted on the fluid by any solid body touching the control volume,

F_B = Force exerted on the fluid body (e.g. gravity),

F_P = Force exerted on the fluid by fluid pressure outside the control volume.

So we say that the total force, F_T, is given by the sum of these forces:

$$F_T = F_R + F_B + F_P$$

The force exerted by the fluid on the solid body touching the control volume is opposite to F_R. So the reaction force, R, is given by:

$$R = -F_R$$

Applications of the Momentum Equation

We will consider the following examples:

- Force due to the flow of fluid round a pipe bend.
- Impact of a jet on a plane surface.
- Force due to flow round a curved vane.

Force due the Flow around a Pipe Bend

Consider a pipe bend with a constant cross section lying in the horizontal plane and turning through an angle of θ°.

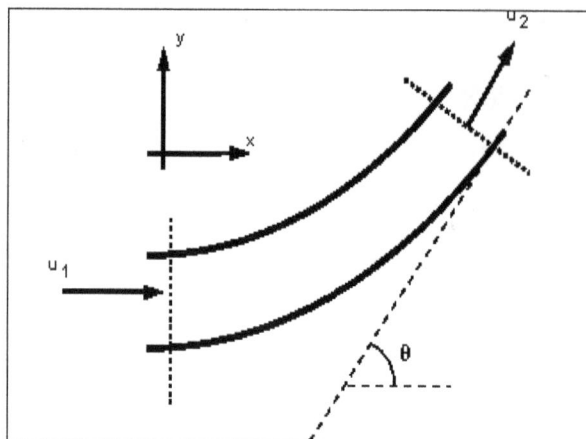

Flow round a pipe bend of constant cross-section.

Why do we want to know the forces here? Because the fluid changes direction, a force (very large in the case of water supply pipes,) will act in the bend. If the bend is not fixed it will move and eventually break at the joints. We need to know how much force a support (thrust block) must withstand.

Step in Analysis:

- Draw a control volume.

- Decide on co-ordinate axis system.

- Calculate the total force.

- Calculate the pressure force.

- Calculate the body force.

- Calculate the resultant force.

Control Volume

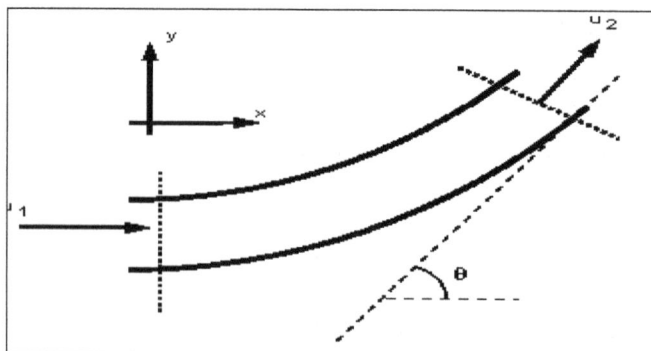

The control volume is draw in the above figure, with faces at the inlet and outlet of the bend and encompassing the pipe walls.

Co-ordinate axis System

It is convenient to choose the co-ordinate axis so that one is pointing in the direction of the inlet velocity. In the above figure the x-axis points in the direction of the inlet velocity.

Calculate the Total Force

In the x-direction:

$$F_{T_x} = \rho Q \left(u_{2_x} - u_{1_x} \right)$$

$$u_{1_x} = u_1$$

$$u_{2_x} = u_2 \cos \theta$$

$$F_{T_x} = \rho Q \left(u_2 \cos \theta - u_1 \right)$$

In the y-direction:

$$F_{T_x} = \rho Q \left(u_{2_y} - u_{1_y} \right)$$

$$u_{1_y} = u_1 \sin 0 = 0$$

$$u_{2_y} = u_2 \sin \theta$$

$$F_{Ty} = \rho Q u_2 \sin \theta$$

Calculate the Pressure Force

$$F_P = \text{pressure force at} 1 - \text{pressure force at } 2$$

$$F_{P_x} = p_1 A_1 \cos 0 - p_2 A_2 \cos \theta = p_1 A_1 - p_2 A_2 \cos \theta$$

$$F_{P_y} = p_1 A_1 \sin 0 - p_2 A_2 \sin \theta = -p_2 A_2 \sin \theta$$

Calculate the Body Force

There are no body forces in the x or y directions. The only body force is that exerted by gravity.

Calculate the Resultant Force

$$F_{T_x} = F_{R_x} + F_{P_x} + F_{B_x}$$

$$F_{T_y} = F_{R_y} + F_{P_y} + F_{B_y}$$

$$F_{R_x} = F_{T_x} - F_{P_x} - 0 = \rho Q \left(u_2 \cos \theta - u_1 \right) - p_1 A_1 + p_2 A_2 \cos \theta$$

$$F_{R_y} = F_{T_y} - F_{P_y} - 0 = \rho Q u_2 \sin \theta + p_2 A_2 \sin \theta$$

And the resultant force on the fluid is given by:

$$F_R = \sqrt{F_{R_x}^2 - F_{R_y}^2}$$

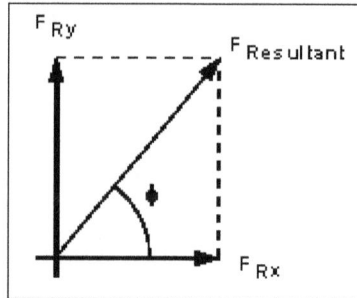

And the direction of application is:

$$\phi = \tan^{-1}\left(\frac{F_{R_y}}{F_{R_x}}\right)$$

the force on the bend is the same magnitude but in the opposite direction:

$$R = -F_R$$

Impact of a Jet on a Plane

We will first consider a jet hitting a flat plate (a plane) at an angle of 90°, as shown in the figure below.

We want to find the reaction force of the plate i.e. the force the plate will have to apply to stay in the same position.

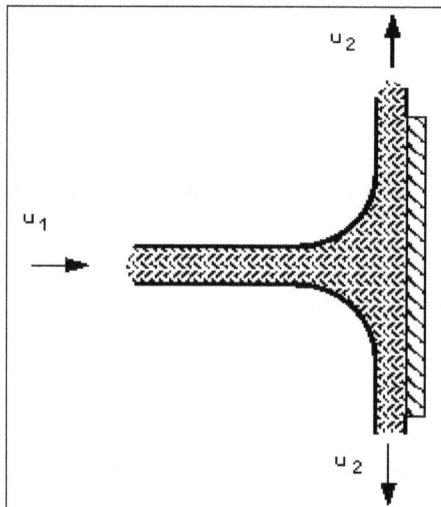

A perpendicular jet hitting a plane.

The analysis take the same procedure:

- Draw a control volume.

- Decide on co-ordinate axis system.

- Calculate the total force.

- Calculate the pressure force.

- Calculate the body force.

- Calculate the resultant force.

The Control volume and Co-ordinate axis are shown in the figure below:

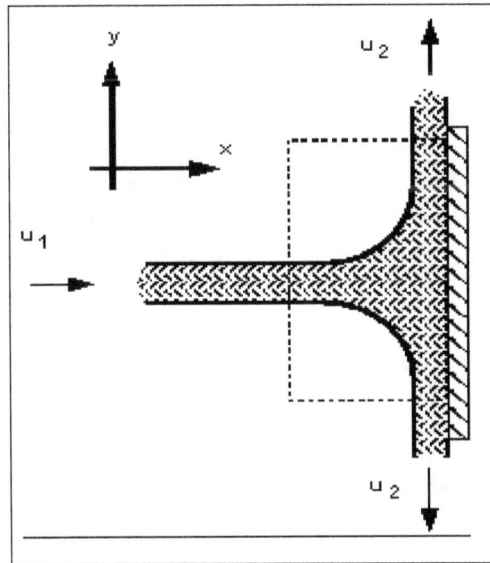

Calculate the Total Force

$$F_{T_x} = \rho Q \left(u_{2_x} - u_{1_x} \right)$$
$$= -\rho Q u_{1_x}$$

As the system is symmetrical the forces in the y-direction cancel i.e.

$$F_{T_y} = 0$$

Calculate the Pressure Force

The pressure force is zero as the pressure at both the inlet and the outlets to the control volume are atmospheric.

Calculate the Body Force

As the control volume is small we can ignore the body force due to the weight of gravity.

Calculate the Resultant Force

$$F_{T_x} = F_{R_x} + F_{P_x} + F_{B_x}$$
$$F_{R_x} = F_{T_x} - 0 - 0$$
$$= -\rho Q u_{1_x}$$

Exerted on the fluid.

The force on the plane is the same magnitude but in the opposite direction:

$$R = -F_{R_x}$$

Force on a Curved Vane

This case is similar to that of a pipe, but the analysis is simpler because the pressures are equal - atmospheric, and both the cross-section and velocities (in the direction of flow) remain constant. The jet, vane and co-ordinate direction are arranged as in the figure below:

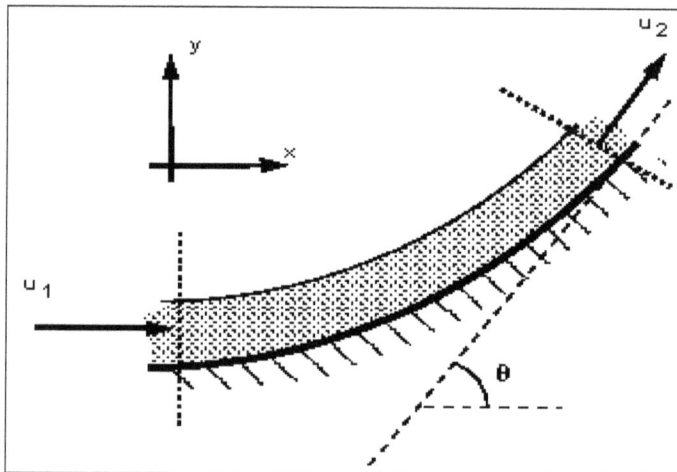

Jet deflected by a curved vane.

1 & 2 Control volume and Co-ordinate axis are shown in the figure above.

Calculate the Total Force in the x-direction

$$F_{T_x} = \rho Q (u_2 - u_1 \cos \theta)$$

But, $u_1 = u_2 = \dfrac{Q}{A}$, so

$$F_{T_x} = -\rho \frac{Q^2}{A}(1 - \cos\theta)$$

And in the y-direction:

$$F_{T_y} = \rho Q (u_2 \sin\theta - 0)$$

$$= \rho \frac{Q^2}{A}$$

Calculate the Pressure Force

Again, the pressure force is zero as the pressure at both the inlet and the outlets to the control volume are atmospheric.

Calculate the Body Force

No body forces in the x-direction, $F_{B_x} = 0$.

In the y-direction the body force acting is the weight of the fluid. If V is the volume of the fluid on he vane then,

$$F_{B_x} = \rho g V$$

(This is often small is the jet volume is small and sometimes ignored in analysis.)

Calculate the Resultant Force

$$F_{T_x} = F_{R_x} + F_{P_x} + F_{B_x}$$
$$F_{R_x} = F_{T_x}$$
$$F_{T_y} = F_{R_y} + F_{P_y} + F_{B_y}$$
$$F_{R_y} = F_{T_y}$$

And the resultant force on the fluid is given by

$$F_R = \sqrt{F_{R_x}^2 - F_{R_y}^2}$$

And the direction of application is exerted on the fluid.

$$\phi = \tan^{-1}\left(\frac{F_{R_y}}{F_{R_x}}\right)$$

The force on the vane is the same magnitude but in the opposite direction.

Energy Equation

For an incompressible flow, where ρ is constant, the primary flow-field variables are P and V.

> Physical principle: Energy can be neither created nor destroyed; It can only change in form.

Consider a fixed amount of matter contained within a closed boundary. This matter defines the system. Because the molecules and atoms within the system are constantly in motion, the system contains a certain amount of energy. For simplicity, let the system contain a unit mass; in turn denote the internal energy per unit mass by e.

The region outside the system defines the surroundings. Let an incremental amount of heat be δq added to the system from the surroundings. Also let δw is the work done on the system by the surroundings.

Both heat and work are forms of energy, and when added to the system, they change the amount of internal energy in the system. Denote this change of internal energy by de. From our physical principle that energy is conserved, we have for the system:

$$\delta q + \delta w = de$$

Equation $\delta q + \delta w = de$ is a statement of the first law of thermodynamics.

Let us apply the first law of fluid flowing through the fixed control volume shown in figure.

> B_1 = rate of heat added to fluid inside control volume from surroundings,
>
> B_2 = rate of work done on fluid inside control volume,
>
> B_3 = rate of change of energy of fluid as it flows through control volume.

From the first law,

$$B_1 + B_2 = B_3$$

First, consider the rate of heat transferred to or from the fluid. This can be visualised as volumetric heating of the fluid inside the control volume due to absorption of radiation originating outside the system or the local emission of radiation by the fluid itself. Let this volumetric rate of heat addition per unit mass be denoted by $\dot{q}$ for the figure the mass contained within an elemental volume is ρdV hence the rate of heat addition to this mass is $\dot{q}(\rho dV)$ summing over the complete control volume. We obtain:

$$\text{Rate of volumetric heating} = \iiint_V \dot{q}\rho dV$$

Let us denote the rate of heat addition to the control volume due to viscous effects. Therefore in equation $B_1 + B_2 = B_3$, the total rate of heat addition is given by equation $= \iiint_V \dot{q}\rho dV$ plus $\dot{Q}_{viscous}$:

$$B_1 = \iiint_V \dot{q}\rho dV + \dot{Q}_{viscous}$$

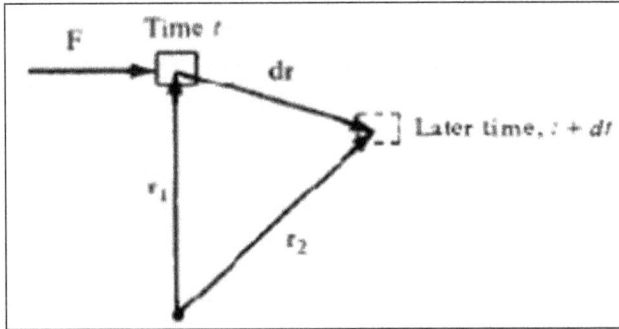

Schematic for the rate of doing work by a force F exerted on a moving body.

Before considering the rate of work done on the fluid inside the control volume, consider a simpler case of solid object in motion, with a force F being exerted on the object as shown in figure. The position of the object is measure from a fixed origin by the radius vector r in moving from r1 to r2 over an interval of time dt.

Rate of doing work on moving body $= \boldsymbol{F} \cdot \boldsymbol{V}$

Rate of work done on fluid inside
V due to pressure force on S
$\quad = -\iint_V (p\,\mathbf{dS}) \cdot \mathbf{V}$

In addition consider an elemental volume dV inside the control volume as shown in figure recalling that f is the body force per unit mass. The rate of work done on the elemental volume due to the body is $(\rho\mathbf{f}\,dV) \cdot \mathbf{V}$.

Summing over the complete control volume, we obtain:

Rate of work done on fluid
inside V due to body forces
$\quad = -\iiint (\rho\mathbf{f}\,dV) \cdot V$

The total rate of work done on the fluid inside the control volume is the sum of equations $-\iint_V (p\,\mathbf{dS}) \cdot \mathbf{V}$, $-\iiint (\rho\mathbf{f}\,dV) \cdot V$ and $\dot{W}_{viscous}$:

$$B_2 = -\iint_S p\mathbf{V} \cdot \mathbf{dS} + \iiint_V \rho(f \cdot \mathbf{V})dV + \dot{W}_{viscous}$$

The elemental flow of total energy across dS is:

$$\left(\rho\mathbf{V}\cdot\mathbf{dS}\right)\left(e+V^2/2\right).$$

We obtain:

Net rate of flow of total energy across control surface $= \oiint_S (\rho\mathbf{V}.\mathbf{dS})\left(e+\dfrac{V^2}{2}\right)$

The total energy inside the complete control volume at any instant in time is:

$$\oiiint_V \rho\left(e+\frac{V^2}{2}\right)dV$$

Therefore,

Time rate change of total energy inside V due to transient variations of flow-field variables $= \dfrac{\partial}{\partial t}\oiiint_V \rho\left(e+\dfrac{V^2}{2}\right)dV$

In turn B_3 is the sum of equations $\oiint_S(\rho\mathbf{V}.\mathbf{dS})\left(e+\dfrac{V^2}{2}\right)$ & $\dfrac{\partial}{\partial t}\oiiint_V \rho\left(e+\dfrac{V^2}{2}\right)dV$:

$$B_3 = \frac{\partial}{\partial t}\oiiint_V \rho\left(e+\frac{V^2}{2}\right)dV + \oiint_S(\rho\mathbf{V}\cdot\mathbf{dS})\left(e+\frac{V^2}{2}\right)$$

Combining the equations:

$$B_1 + B_2 = B_3, \quad B_1 = \oiiint_V \dot{q}\rho\, dV + \dot{Q}_{viscous}, \quad \oiint_S(\rho\mathbf{V}.\mathbf{dS})\left(e+\frac{V^2}{2}\right),$$

and

$$B_3 = \frac{\partial}{\partial t}\oiiint_V \rho\left(e+\frac{V^2}{2}\right)dV + \oiint_S(\rho\mathbf{V}\cdot\mathbf{dS})\left(e+\frac{V^2}{2}\right)$$

We get:

$$\oiiint_V \dot{q}\rho\, dV + \dot{Q}_{viscous} - \oiint_S p\mathbf{V}\cdot\mathbf{dS} + \oiiint_V \rho(\mathbf{f}\cdot\mathbf{V})dV + \dot{W}_{viscous}$$

$$= \frac{\partial}{\partial t}\oiiint_V \rho\left(e+\frac{V^2}{2}\right)dV + \oiint_S \rho\left(e+\frac{V^2}{2}\right)\mathbf{V}\cdot\mathbf{dS}$$

The above equation represents the energy equation in integral form.

Applying the divergence theorem to the surface integrals in above equation collecting all terms inside the same volume integral, and setting the integrand equal to zero, we obtain

$$\frac{\partial}{\partial t}\left[\rho\left(e+\frac{V^2}{2}\right)\right]+\nabla\cdot\left[\rho\left(e+\frac{V^2}{V}\right)\mathbf{V}\right]=\rho\dot{q}-\nabla\cdot(p\mathbf{V})+\rho(\mathbf{f}\cdot\mathbf{V})+\dot{Q}'_{viscous}+\dot{W}'_{viscous}$$

If the flow is steady, inviscid, adiabatic, without body forces the equation:

$$\oiiint_V \dot{q}odV+\dot{Q}_{viscous}-\oiint_S p\mathbf{V}\cdot\mathbf{dS}+\oiiint_V \rho(\mathbf{f}\cdot\mathbf{V})dV+\dot{W}_{viscous}$$

$$=\frac{\partial}{\partial t}\oiiint_V \rho\left(e+\frac{V^2}{2}\right)dV+\oiint_S \rho\left(e+\frac{V^2}{2}\right)\mathbf{V}\cdot\mathbf{dS}$$

and

$$\frac{\partial}{\partial t}\left[\rho\left(e+\frac{V^2}{2}\right)\right]+\nabla\cdot\left[\rho\left(e+\frac{V^2}{V}\right)\mathbf{V}\right]=\rho\dot{q}-\nabla\cdot(p\mathbf{V})+\rho(\mathbf{f}\cdot\mathbf{V})+\dot{Q}'_{viscous}+\dot{W}'_{viscous}$$

reduces to,

$$\oiint_S \rho\left(e+\frac{V^2}{2}\right)\mathbf{V}\cdot\mathbf{dS}=-\oiint_S p\mathbf{V}\cdot\mathbf{dS}$$

and

$$\nabla\cdot\left[\rho\left(e+\frac{V^2 2}{2}\right)\mathbf{V}\right]=-\nabla\cdot(\rho\mathbf{V})$$

If the gas is calorically perfect, then:

$$e=c_vT$$

The system can be completed by using the perfect gas equation of state:

$$p=\rho RT$$

References

- Motion, physics: physics-and-radio-electronics.com, Retrieved 14 April, 2019
- Speed-2699009: thoughtco.com, Retrieved 16 July, 2019

- Velocity-definition-in-physics-2699021: thoughtco.com, Retrieved 23 January, 2019

- Aerodynamic-force: encyclopedia2.thefreedictionary.com, Retrieved 25 March, 2019

- What-is-momentum-2698743: thoughtco.com, Retrieved 18 June, 2019

- Centre-of-Pressure: skybrary.aero, Retrieved 25 April, 2019

- Airspeed, training-and-safety-students-maneuvers-topics: aopa.org, Retrieved 17 July, 2019

Aeroelasticity: A Comprehensive Study

The branch of physics that deals with the study of interaction between the inertial, elastic, and aerodynamic forces is defined as aeroelasticity. A few of its types are static aeroelasticity, dynamic aeroelasticity, nonlinear aeroelasticity, etc. This chapter discusses these types of aeroelasticity in detail.

Aeroelasticity is a division of applied mechanics, which analyzes interaction of the aircraft as an elastic system (or elastic aerial vehicle) with atmosphere. Aeroelastic phenomena appear in many engineering areas, for example, structural engineering when studying the wind effect on the bridges and skyscrapers, or in marine application and power-plant engineering. Aeroelasticity analysis is especially important in aviation and rocket science.

Aerodynamic forces acting on the aircraft during the flight cause deformations of the elastic structure, which in turn lead to variation of aerodynamic forces.

All the phenomena considered within the scope of aeroelasticity are divided into static and dynamic. The phenomena that involve interaction of aerodynamic forces and elastic forces of the structure are classified as static. Divergence of the lifting surfaces (i.e. wing, fin), reverse of the aircraft controllers, effect of the structure elasticity on redistribution of aerodynamic pressure and on static aircraft stability are static phenomena considered within the scope of aeroelasticity. The phenomena, for which interaction of aerodynamic inertial forces with elastic forces becomes critical, are classified as dynamic. Flutter, stalling flutter, buffeting, transonic self-oscillations of the aircraft controllers, and response of the elastic structure on the aircraft dynamic stability are dynamic phenomena considered within the scope of aeroelasticity.

Aeroelasticity concerns the interaction between aerodynamics, dynamics and elasticity. This interaction can result in negatively or badly damped wind turbine blade modes, which can have a significant effect on the turbine lifetime. The first aeroelastic problem that occurred on commercial wind turbines concerned a negatively damped edgewise mode. It is important to ensure that there is some out-of-plane deformation in this mode shape to prevent the instability. For larger turbine blades with lower torsional stiffness and the possibility of higher tip speeds for the offshore designs, classical flutter could also become relevant. When designing a wind turbine blade, it is therefore crucial that there is enough damping for the different modes and that there is no coincidence of natural frequencies with excitation frequencies (resonance). An effective aeroelastic analysis is also important, and the tools used for such an analysis must include the necessary detail in the structural model.

Static Aeroelasticity

Static aeroelasticity is the study of the deflection of flexible aircraft structures under aerodynamic loads, where the forces and motions are considered to be independent of time. Consider the aerodynamic lift and moment acting upon a wing to depend solely upon the incidence of each chord wise strip (i.e. strip theory). These loads cause the wing to bend and twist, so changing the incidence and consequently the aerodynamic flow, which in turn changes the loads acting on the wing and the deflections, and so on until an equilibrium condition is usually reached. The interaction between the wing structural deflections and the aerodynamic loads determines the wing bending and twist at each flight condition, and must be considered in order to model the static aeroelastic behavior. The static aeroelastic deformations are important as they govern the loads in the steady flight condition, the lift distribution, the drag forces, the effectiveness of the control surfaces, the aircraft trim behavior and also the static stability and control characteristics. The aeroelastic wing shape at the cruise condition is of particular importance as this has a crucial effect on the drag and therefore the range.

Most of the modern aircrafts have high-aspect-ratio swept wing and use composite material, which makes the static aeroelasticity become more and more severe. For instance, the wing on the Boeing 787 Dreamliner will nominally deflect 10 feet at cruise. For the high-aspect-ratio swept wing, it will give rise to the wing deflection and torsion because of the static aeroelasticity problems, which will cause the angle of attack at the local section becomes smaller, and changes the distribution of surface pressure. For the transonic aircraft, the decrease of angle of attack at the local section will also influence the intensity of shock wave. Therefore, it is crucial to analyze the static aeroelasticity for modern aircrafts.

By CFD/CSD method, loosely coupling is usually utilized to increase the computational efficiency. Hence, we can analyze the structure and the flow field separately. When it comes to structural analysis, there are two methods: one is linear analysis method. The wing deflection can be depicted by the modal method, under the assumption of small deflection. The other is nonlinear method. Finite element analysis is usually applied. Mian et al. used this method to analyze the nonlinear deflection of high-aspect-ratio wing. In addition, both the multibody method and nonlinear aeroelastic scaling method are also used. Here we used the computational structural dynamics (CSD) to analyze the structural deflection. On the other hand, when it comes to aerodynamic analysis, there are also two methods, linear and nonlinear methods, which are similar to structural analysis methods. Traditional aerodynamic linear theory is usually applied in the linear analysis method. For certain conditions, linear theory works well. However, the conditions are limited to the subsonic and supersonic flows. When it refers to nonlinear aerodynamic calculation in transonic flow, computational fluid dynamics (CFD) is usually applied at present. When it comes to the static aeroelastic analysis, the steady CFD solution of the rigid wing is calculated firstly. Next, we get the generalized force and calculate the deflection of wing to obtain the new boundary by coupling with the static equilibrium equation. After that, we change the wall boundary by grid deformation method and use CFD to calculate the aerodynamic force on the deformed wing. The results can be obtained by repeating the above steps in order, until the deflection converges or diverges. However, when it comes to the complex three-dimensional configurations, this method is time-consuming and has high computational complexity.

More and more researchers now used the surrogate method in their studies, as the problems mentioned above exist in the fluid-structure coupling problems and optimum design. Surrogate methods can fit the nonlinear multiple-input/output function accurately and has high computational efficiency. Due to these advantages, more and more researchers utilized these methods to model the nonlinear unsteady aerodynamics. Lindhorst et al. combined the parameter reduction via proper orthogonal decomposition and system identification methods to model nonlinear unsteady two-dimensional aerodynamics. And the model can accurately predict the static and transient response of the airfoil. Furthermore, they demonstrated the application on a three-dimensional case, the high-Reynolds-number aerostructural dynamics (HIRENASD), and the model can capture the influences of nonlinear aerodynamic effects on the forces. Moreover, the model can be used in both static and transient aeroelastic investigations at a fixed Mach number. Kou and Zhang applied radial basis function neural network to model two-dimensional nonlinear aerodynamics. And the approach can capture both linear and nonlinear characteristic.

CFD and CSD Coupling Method

The flow governing equations used to solve the aerodynamics can be written as:

$$\frac{\partial}{\partial t}\iiint_{\Omega} \mathbf{Q}dV + \iint_{\partial\Omega} \mathbf{F}(\mathbf{Q}) \cdot \mathbf{n}dS = \iint_{\partial\Omega} \mathbf{G}(\mathbf{Q}) \cdot \mathbf{n}dS,$$

where Ω is the control volume, $\partial\Omega$ is the boundary of the control volume, $\mathbf{n}$ is the outer normal vector of the control volume boundary, V denotes the volume of the element, and S denotes the surface area of each surface. When it comes to two dimension, V denotes the surface area, and S denotes the length. $\mathbf{Q}$ is the vector of conservative variables, $\mathbf{F}(\mathbf{Q})$ is the vector of the inviscid fluxes, and $\mathbf{G}(\mathbf{Q})$ denotes viscous fluxes.

The CFD solver based on the steady Reynolds-Averaged Navier-Stoke (RANS) equations has the ability to simulate the flow with viscous effects. The Spalart-Allmaras (S-A) turbulence model works well in describing the viscosity in the transonic flow, in which shock wave exists.

Under the assumption of linear small deflection, the modal method is utilized to describe the wing deflection. Since there is no need to take the structural inertial force into consideration, the deflection is computed by the static equilibrium equation:

$$\mathbf{K}\xi = \mathbf{F},$$

where ξ is the generalized displacement vector, $\mathbf{K}$ is the generalized stiffness matrix, and $\mathbf{F}$ is the generalized force vector.

To get rid of the dynamic pressure effect, the generalized forces are nondimensionalized by dynamic pressure, which are called as generalized force coefficients:

$$\mathbf{f} = \frac{\mathbf{F}}{Q},$$

where $\mathbf{f}$ represents the generalized force coefficient vector and Q represents the dynamic pressure.

The new structural boundary can be depicted by the following equation:

$$\mathbf{x}_{new}^{j+1} = \eta \cdot \mathbf{x}_{new}^{j} + (1-\eta) \cdot \mathbf{x}_{old}^{j},$$

where $\mathbf{x}_{new}^{j+1}$ represents the coordinate matrix of the boundary grid nodes at the $j+1$th iteration and $\mathbf{x}_{old}^{j}$ represents the coordinate matrix of the boundary grid nodes at the jth iteration. $\mathbf{x}_{new}^{j}$ represents the coordinate matrix of the grid nodes, which is obtained from the new wing deflection. It is calculated according to the generalized forces of the jth iteration $(j = 0,1,2,.....)$ and $\eta \in (0,1,0)$.

The grid is deformed according to the new boundary by spring analogy method. In addition, loosely coupling is usually utilized to increase the computational efficiency.

Kriging Model

The Kriging surrogate model is a kind of model aimed at minimizing variance and constructing an unbiased estimation of the spatial distribution data via the statistical method of stochastic process. The functional expression can be given as:

$$y(x) = P(\beta, x) + z(x) = \mathbf{f}^{T}(x)\beta + z(x),$$

where, $P(\beta, x)$ is the regression model and β is the regression parameters. $z(x)$ is the nonparametric random function, and its statistical properties are written as.

Mean value:

$$E\big[z(x)\big] = 0$$

Variance:

$$Var\big[z(x)\big] = \sigma^{2},$$

Covariance:

$$Cov\big[z(x_{i}), z(x_{j})\big] = \sigma^{2}\big[R_{ij}(\theta, x_{i}, x_{j})\big],$$

where, x_i and x_j are the design sites, and $R_{ij}\left(\theta, x_i, x_j\right)$ is the function with parameter θ and represents the spatial relativity among the design sites. The spatial relativity between every two design sites is related to their spatial distance. Hence, it can be depicted by the following equation:

$$R_{ij}\left(\theta, x_i, x_j\right) = \prod_{k=1}^{n} R_k\left(\theta_k, d_k\right),$$

where n is the number of design variables and d_k is the distance between every two design sites. The concrete function is given as:

$$d_k = \left|x_i^k - x_j^k\right|, i, j = 1, 2, \ldots n,$$

where x_i^k and x_j^k are the coordinate values of the ith and jth design sites in the kth direction and θ_k is the constant parameter of the function in the kth direction.

Aimed to minimize $\sigma^2\left(x_{new}^*\right)$, after the mathematical derivation, the predictor is computed as:

$$\hat{y}\left(x_{new}^*\right) = f\left(x_{new}^*\right)^T \beta^* + r\left(x_{new}^*\right)^T \gamma^*,$$

where, $\beta^* = \left(P^T R^{-1} P\right)^{-1} P^T R^{-1} Y$ and $R\gamma^* = Y - P\beta^*$ where $P = \left[p_1, p_2, \ldots\ldots p_n\right]^T$, which represents the vector of regression values for the design sites, and Y is the response array of the design sites. Since β^* and γ^* are related to the design sites instead of the predictor sites and the predictor sites are only related to $f\left(x_{new}^*\right)$ and $r\left(x_{new}^*\right)$, the predicted response $\hat{y}\left(x_{new}^*\right)$ will soon be obtained when" x_{new}^* is given.

We utilized the Gauss Function as **R** matrix:

$$\mathbf{R}_k\left(\theta_k, d_k\right) = \exp\left(-\theta_k d_k^2\right)$$

where $\theta_k = 0.1$.

The Static Aeroelastic Analysis Method

Sampling Method

As the model is expected to calculate the generalized force coefficients at the different conditions of Mach numbers (Ma), angles of attack (AOA), and dynamic pressures (Q), model samples need to be chosen at three steps.

The first step is to choose some sets of Ma and AOA from a certain range by Latin hypercube sampling (LHS). The nonlinear effect in transonic flow occurs when either MA or AOA changes. For example, the aerodynamic force would change nonlinearly even if either AOA or Ma increases linearly due to the effect of the shock wave. Hence, the nonlinear effect of the Ma and AOA should be taken into consideration. The next step is to choose several sets of dynamic pressures at each set Ma of AOA and. Then, the static aeroelasticity will be analyzed at each condition to obtain the corresponding equilibrium position. The third step is to choose several sets of generalized displacements by LHS from a certain range close to each equilibrium position, in order to ensure that the numerical value of each mode generalized displacement is limited to a certain range. The obtained generalized displacements need to be able to describe the real wing deformations and enable the deformations to change in a certain range, since the model needs to predict the aerodynamic forces for the wing of different deformations. This sampling method can satisfy the requirements, so we gained the generalized displacements in this way. Finally, CFD method is used to calculate the corresponding generalized force coefficients and aerodynamic force coefficients at each training case. Hence, the sum of training cases can be calculated by the following function:

$$S_{total} = sampling_{Ma,AOA} \times sampling_Q \times sampling_\xi,$$

where, S_{total} represents the sum of training cases, $sampling_{Ma,AOA}$ represents the sampling sets of Ma and AOA $sampling_Q$ represents the sampling sets of Q, and $sampling_\xi$ represents the sampling sets of generalized displacements.

Modeling Method

A model is required to calculate the generalized forces at different Ma, AOA, and deflections. Hence, besides a set of generalized displacements, Ma and AOA will also be input to the model. And the output is a set of the corresponding generalized force coefficients. The relation between the inputs and the output is given as:

$$f = f_{gf}\left(Ma, \ AOA, \xi\right)$$

where f is the vector of generalized force coefficients and ξ is the vector of generalized displacements. To get the generalized forces, we need to multiply the f by the dynamic pressure, for the coefficients are nondimensionalized by the dynamic pressure. The forces would be used to calculate the deformation of wing, which will put forward the procedure. This model is utilized to replace the CFD flow solver and will be called many times during the static aeroelastic analysis.

Dynamic Aeroelasticity

Dynamic aeroelasticity studies the interactions among aerodynamic, elastic, and inertial forces. Examples of dynamic aeroelastic phenomena are:

Flutter

Flutter is a dynamic instability of an elastic structure in a fluid flow, caused by positive feedback between the body's deflection and the force exerted by the fluid flow. In a linear system, "flutter point" is the point at which the structure is undergoing simple harmonic motion—zero net damping—and so any further decrease in net damping will result in a self-oscillation and eventual failure. "Net damping" can be understood as the sum of the structure's natural positive damping and the negative damping of the aerodynamic force. Flutter can be classified into two types: *hard flutter*, in which the net damping decreases very suddenly, very close to the flutter point; and *soft flutter*, in which the net damping decreases gradually.

In water the mass ratio of the pitch inertia of the foil to that of the circumscribing cylinder of fluid is generally too low for binary flutter to occur, as shown by explicit solution of the simplest pitch and heave flutter stability determinant.

Structures exposed to aerodynamic forces—including wings and aerofoils, but also chimneys and bridges—are designed carefully within known parameters to avoid flutter. In complex structures where both the aerodynamics and the mechanical properties of the structure are not fully understood, flutter can be discounted only through detailed testing. Even changing the mass distribution of an aircraft or the stiffness of one component can induce flutter in an apparently unrelated aerodynamic component. At its mildest, this can appear as a "buzz" in the aircraft structure, but at its most violent, it can develop uncontrollably with great speed and cause serious damage to or lead to the destruction of the aircraft, as in Braniff Flight 542, or the prototypes for the VL Myrsky fighter aircraft. Famously, the original Tacoma Narrows Bridge was destroyed as a result of aeroelastic fluttering.

Aeroservoelasticity

In some cases, automatic control systems have been demonstrated to help prevent or limit flutter-related structural vibration.

Propeller Whirl Flutter

Propeller whirl flutter is a special case of flutter involving the aerodynamic and inertial effects of a rotating propeller and the stiffness of the supporting nacelle structure. Dynamic instability can occur involving pitch and yaw degrees of freedom of the propeller and the engine supports leading to an unstable precession of the propeller. Failure of the engine supports led to whirl flutter occurring on two Lockheed L-188 Electra in 1959 on Braniff Flight 542 and again in 1960 on Northwest Orient Airlines Flight 710.

Transonic Aeroelasticity

Flow is highly non-linear in the transonic regime, dominated by moving shock waves. It is mission-critical for aircraft that fly through transonic Mach numbers. The role of shock waves was first analyzed by Holt Ashley. A phenomenon that impacts stability of aircraft known as "transonic dip", in which the flutter speed can get close to flight speed, was reported in May 1976 by Farmer and Hanson of the Langley Research Center.

Buffeting

Buffeting is a high-frequency instability, caused by airflow separation or shock wave oscillations from one object striking another. It is caused by a sudden impulse of load increasing. It is a random forced vibration. Generally it affects the tail unit of the aircraft structure due to air flow downstream of the wing.

The methods for buffet detection are:

- Pressure coefficient diagram,

- Pressure divergence at trailing edge,

- Computing separation from trailing edge based on Mach number,

- Normal force fluctuating divergence.

Nonlinear Aeroelasticity

Nonlinearities in aeroelasticity include structural nonlinearities and aerodynamic nonlinearities. In some situations, such as shock waves in transonic flows, wing tip vortices, and dynamic stall, aerodynamic nonlinearities have to be taken into account. Nonlinear aerodynamic effects are more difficult to analyse because the fluid motion is governed by equations where analytical solutions are practically non-existent. A fully nonlinear aerodynamic code to solve the Euler equations coupled with a structural model in a two-dimensional flow case has been developed by Djayapertapa and Allen and Djayapertapa et al. However, the use of full computational fluid dynamic (CFD) and computational structural dynamics (CSD) coupled codes in design loops is very time-consuming, particularly when considering an active control system. Additionally, the lack of visibility of the full nonlinear equations in state space form makes various forms of control law design and stability analysis extremely difficult if not impossible. A reduced order model (ROM) of the Euler code was then created, and the full nonlinear and reduced order aerodynamic models in control law design were compared. With an equivalent accuracy to the CFD method, the ROM requires a computational time that is much more comparable with traditional linear methods.

Conversely, structural nonlinearities could arise from worn hinges of control surfaces, loose control linkages, and material behavior as well as various other sources. Aging and combat aircraft that carry heavy external stores are more concerned with the effects associated with nonlinear structures. Structural nonlinearities may be classified as being either distributed or concentrated. In general, distributed structural nonlinearities are governed by elasto dynamic deformations that affect the whole structure. Concentrated nonlinearities, on the other hand, act locally and are commonly found in control mechanisms or in the connecting parts between wing, pylon, engine or external stores. The concentrated nonlinearities can be classified basically into three types: cubic nonlinearity, freeplay nonlinearity, and hysteresis nonlinearity. Time delay is another nonlinearity that should be considered in the problem of aeroelastic control. The three classical structural nonlinearities have been investigated by many researchers, but they are not exactly the same with the relationships from experimental test.

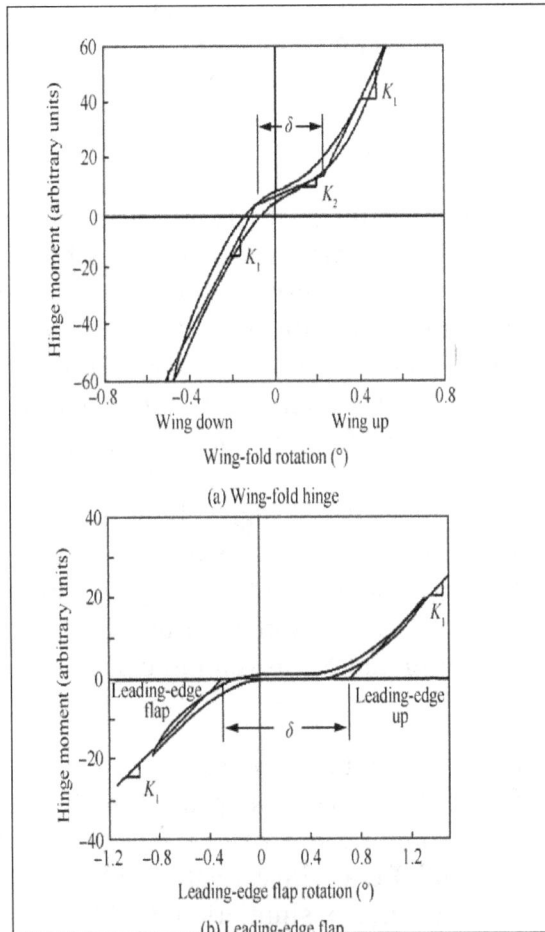

(a) Wing-fold hinge

(b) Leading-edge flap

Nonlinearities from testing data.

Two examples of the hysteresis nonlinear relationship from ground testing data are shown in figure, K_1 and K_2 are the stiffness of spring, δ is the amount of freeplay. But they were ignored and simplified as bilinear and freeplay nonlinearities in their investigation. With the unsteady aerodynamic model from vortex lattices formulated in a continuous time domain and expressed in a dimensionless form, a combined nonlinearity of freeplay and cubic stiffening was investigated by Zhao and Hu. In order to improve the nonlinear model, rational polynomial approximation could be used to describe the nonlinear relationship. And for hysteresis nonlinearity, the item of velocity has to be included. The work of Li et al. showed that the aeroelastic system with rational polynomials in pitch exhibits almost exactly the same dynamic responses, such as convergence and LCO, with what were seen in the system with freeplay or hysteresis nonlinearities. The theoretical analysis of the energy transformation shows that the switching point in the bilinear or hysteresis nonlinearity has no effect on the aeroelastic response of the system, which was also verified by numerical examples. Nonlinear relationship in aeroelastic system should be tested from the actual aircraft structures. Until now, there is not enough data from experimental test. Rational polynomial model has advantages to describe the nonlinear relationship in aeronautical structures. And the aeroelastic system with rational polynomial is continuous and derivable, so it can be solved by using various methods including theoretical approaches.

Aeroelastic Control

For the two-dimensional wing section with structural nonlinearity, research effort has been made to develop control strategies to suppress flutter. In the early stage, the classical linear full-state feedback control law was derived for a wing section with nonlinear stiffness to stabilize the nonlinear system in some circumstances. Partial feedback linearization methodology was also applied to the design of nonlinear controllers for nonlinear aeroelastic system. In order to derive a globally stabilizing controller, a full feedback linearization controller based on two control surfaces was designed. The state-dependent Riccati equation (SDRE) method was developed for nonlinear control problems, and used to design suboptimal control laws of nonlinear aeroelastic systems considering both quasi-steady and unsteady aerodynamics 36. Based on the SDRE control law, the effect of freeplay and time delay in the control surface of a closed-loop system was investigated by Li et al. Time delay feedback was successfully used by Ramesh and Narayanan to control the chaotic motions in a two-dimensional airfoil.

When the uncertainty in the structural nonlinearity was taken into account, an adaptive control method would be used to depress the aeroelastic flutter. General sources of uncertainty that complicate airframe design and testing were briefly described by Pettit. Pettit and Beran investigated the effects of uncertainty on airfoil LCO by using of Monte Carlo simulation (MCS). Parametric uncertainty was modeled in the third- and fifth-order stiffness coefficients of the pitch spring. Different computational methodologies, such as Wiener-Haar, Cyclic and B-spline projection methods have been developed to quantify the uncertain response of an airfoil aeroelastic system in limit-cycle oscillation, subject to parametric variability. Uncertainties are specified in the cubic coefficient of the torsional spring and in the initial pitch angle of the airfoil. When the uncertainty was considered in the flutter suppression, adaptive controllers based on partial or full feedback linearization were derived. A series of adaptive controllers was derived for flutter suppression by Singh and Brenner, and unstructured uncertainties were also taken into account. Recently, an output feedback and an adaptive decoupled fuzzy sliding-mode control laws have been implemented for suppressing flutter and reducing the vibrational level in sub-critical flight speed range. An ultrasonic motor was also used for the flutter control of the aeroelastic system with a nonlinear stiffness in pitch. Structured model reference (SMR) adaptive control method has been developed for a special type of structure, and used for flutter suppression of an aeroelastic system.

Damping uncertainty in airframe structure and control system is inevitable and may have significant effect on the aeroelastic behavior. It is very difficult to establish an accurate damping model, and much of experimental data is normally needed. Two examples of the wing sections with single trailing-edge or leading- and trailing-edge were taken. Adaptive controllers based on partial feedback linearization and structured-model reference were designed. The numerical simulation results show that the damping uncertainty has a positive effect on the control effectiveness. The closed-loop system considering damping uncertainty has quicker response to control and greater critical flutter velocity.

Nonlinear Aeroelasticity of High-aspect-ratio Wings

As shown in figure below, high-aspect-ratio flexible wings exhibits large deformation under aerodynamic loads, and geometric nonlinearity has to be taken into account in structural modeling. For the aeroelastic analysis of high-aspect-ratio wings, linear or nonlinear unsteady aerodynamics needs to be considered. Presently, nonlinear beam model and two-dimensional strip theory have been extensively favored in identifying the critical nonlinear aeroelastic phenomena of flexible structure.

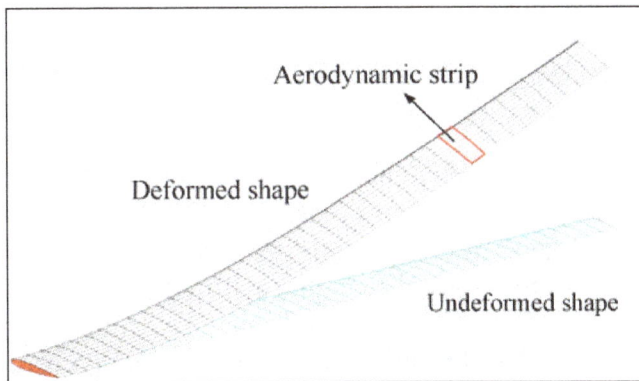

Deformation of high-aspect-ratio flexible wing.

Structural Modeling

With the application of stronger, lighter weight and more flexible synthetic materials, high-aspect ratio wings may suffer from large deformation, which makes geometrically nonlinear analysis of structure become essential. Generally, geometric nonlinearity can be categorized into three types: large displacement/large rotation/small strain, large displacement/small rotation/small strain and large displacement/large rotation/large strain, among which the "large displacement/large rotation/small strain" geometric nonlinear behavior is commonly seen in HALE aircraft and helicopter blades. On hypothesis of small strain, the structure model of high-aspect-ratio flexible wing is usually represented as a beam considering geometric nonlinearity, and two types of fundamental structural models are widely used, the nonlinear moving beam and the geometrically exact intrinsic beam.

The nonlinear moving beam model has the feature of describing the beam deformed position with displacements or Euler angles. The initial form was presented by Hodges and Dowell 57 for rotor blades using Hamilton's principle and Newtonian method in 1974. Their model coupled bending and torsion deflection with second-order nonlinear terms and used displacements to describe the beam deformed position. Then, some researchers improved the nonlinear beam model by involving more nonlinear terms so as to describe deformation more accurately However, the results show that selection of higher-order nonlinear terms could lead to contradiction between the computational complexity and displacement accuracy. Later Minguet and Dugundji used the Euler angles to describe the beam deformed position and derived equilibrium and compatibility equations without any limitation on magnitude of displacements. But the Euler angles somewhat make the aerodynamic loads calculation a bit complex. The nonlinear moving beam model has great influence on structural modeling of high-aspect-ratio flexible wing, even though it was originally developed for rotor aeroelastic analysis with some imperfection. Jaworski and Dowell 60 conducted the comparison of theoretical structural models with experiment for high-aspect-ratio wing. Liu and Xiang studied the nonlinear flutter characteristics of high-aspect-ratio flexible wing. Pan et al. investigated the flight loads analysis and optimization.

The geometrically exact intrinsic beam model was mainly completed by Hodges and his co-workers.

The model provides a geometrically exact, fully intrinsic dynamic formulation for moving beam including initial curvature and twist, shear deformation, rotary inertia, and general anisotropy. The governing equations are characterized by partial differential equations of motion and kinematical formulas with unknowns only including stress resultants, generalized strains and generalized velocities, and appear to be unique for the absence of displacement and rotation variables. The main advantage of the theory is that the governing equations are geometrically exact and written in intrinsic form without being augmented with some forms of angular displacement variables, and with the degree of nonlinearity no higher than two, which simplifies derivation of aeroelastic equations and reduces computational cost. It has already been applied to structural modeling for high-aspect-ratio flexible wing aeroelastic analysis by some researchers and would have a wide application prospect in structural modeling. Chang and Hodges presented parametric studies on ground vibration test of HALE aircraft. Zhang and Xiang studied the nonlinear aeroelastic response of flexible wing and the parametric effect when subjected to a lateral follower force.

Aerodynamic Modelling

The key for a successful aeroelasticity analysis is the calculation of the aerodynamic loads based on the motion/deformation of the aerodynamic body. For high-aspect-ratio flexible wing, aerodynamic loads calculation can be formulated based on different types of aerodynamic theory, such as strip theory, unsteady vortex-lattice method (UVLM), indicial response theory and Euler/Navier-Stokes CFD aerodynamic modelling techniques, while the strip theory has been widely applied for its higher computational efficiency. Aerodynamic loads in strip theory are described by airfoil aerodynamic model, which can be divided into two major catalogues: frequency domain analytical models and time domain numerical models.

Frequency domain analytical models mainly include Greenberg's aerodynamic model, Loewy's aerodynamic model, Theodorsen aerodynamic model and so on. In frequency domain analytical models, the motion of airfoil and airflow are assumed to be harmonic functions of time, and the equation of wing coupling with nonlinear structure and aerodynamic model can be formulated in the frequency domain. The Frequency domain analytical models can easily deduce the aeroelastic problems into an eigenvalue problem, and the flutter boundaries and stability analysis can be determined by eigenvalues analysis. It is equivalent in aeroelastic stability analysis between the treatment of time and frequency domains. However, frequency domain analytical models, strictly speaking, are only effective when parameters are close to flutter boundaries. It is inaccurate for transient aeroelastic analysis while the motion is far away from flutter boundaries.

Time domain numerical models can obtain airfoil aerodynamic loads in arbitrary motion, and are widely used in nonlinear aeroelastic analysis. Currently, the most commonly used time domain numerical models include unsteady wake models and dynamic stall models. In unsteady wake models, the unsteady aerodynamic effects are considered as a consequence of the time-history of the induced velocity from the vorticity contained in the shed wake, coupled with the induced velocity contributed by the circulation contained in the trailed wake. There are two general approaches to model the induced effects produced by this cycloidal wake: dynamic inflow models and vortex wake models. A present, the most popular model of dynamic inflow theory is that of Pitt and Peters. The dynamic inflow models have attractive mathematical forms and relatively high numerical efficiency, which will always be appealing for high-aspect-ratio flexible wing aeroelastic analysis. However, the concept of the apparent mass applied to the time constants seems certainly not to be a rigorous analogy. The vortex wake models use "vortex methods" to represent the strengths (circulation) and spatial locations of the vertical elements and mainly contain the prescribed vortex method and the free vortex method. The disadvantage of the vortex wake models is the relatively higher computational cost.

Dynamic stall will occur on any airfoil or lifting surface when the effective angle of attack is above its normal static stall angle. This phenomenon has been extensively studied using oscillating 2-D airfoils in wing tunnel experiments. Based on experimental data, several dynamic stall models have been provided using parsimonious, semi-empirical formulas. A few of them are currently used for aeroelastic analysis, such as ONEAR model and Leishman-Beddoes model. The ONERA dynamic stall model was first developed by Tran and Petot. They described the unsteady airfoil behavior in both attached flow and separated flow of a pitching airfoil using a set of nonlinear differential equations. Then some suggested modifications were added and a later common used version of ONERA model was documented by Petot. The coefficients in these differential equations of the ONERA model are determined by parameter identification using experimental measurements on oscillating airfoils. At present, ONERA aerodynamic model is widely applied to aeroelastic analysis by many researchers such as Zhao and Hu and Zhang and Xiang. However the ONERA models usually lack rigor and generality and need a significant number of empirical coefficients. Later Troung presented ONERA-BH model using a Vander Pol Duffing type nonlinear equation to represent the separated flow conditions. The model requires slightly fewer coefficients, and can give good description of the phenomenon of vortex exfoliation. The Leishman-Beddoes (L-B) model was first presented by Beddoes, and then developed by Leishman et al. The L-B model includes static model, unsteady attached flow model, trailing edge separated flow model and vertex flow model, and has the ability of representing the unsteady lift, pitching moment and drag characteristics of an airfoil undergoing dynamic stall. An important feature of this model is that rigorous representations of compressibility effects are included, which are essential for helicopter applications. Although the model has also been developed as a set of differential equations and has fewer coefficients, it shows significant disagreement with the experimental data at low Mach number and has yet seldom been seen in high-aspect-ratio flexible wing aeroelastic analysis.

The strip theory coupled with two-dimensional airfoil aerodynamic model has been widely used for aerodynamic loads calculation as that it easily allows for corrections, including semi-empirical stall models, and steady viscous drag. However, it shows disadvantage in evaluation of spanwise variations, which may be critical for high-aspect-ratio wings when large deflections occur. Therefore, the approach of nonlinear beam model coupled with strip theory could cause relevant 3-D flow physics to be neglected, such as the accurate prediction of wing-tip effects, and the aerodynamic interference between wakes and lifting surfaces. Currently, Euler/Navier-Stokes equations and CFD techniques have still been used to simulate the 3-D flow of flexible aircraft, but incur computational expenses. Some researchers point out that the UVLM, which provides a medium-fidelity tool for aerodynamic calculation, may be unveiled as an outstanding one in future aeronautical research.

Nonlinear Aeroelasticity of Aircraft

Nonlinear aeroelastic phenomena are commonly seen and becoming increasingly important in the aeroelastic analysis of the full aircraft, especially for HALE aircraft and fight aircraft.

HALE Aircraft

For the HALE aircraft with flexible wing, as shown in figure, the frequency of elastic mode is quite low, which appears near to the frequency of flight mode, and the coupling effect of aeroelastic behavior and flight dynamics occurs known as body-freedom flutter. The phugoid mode is mildly unstable, and the structural dynamics and the rigid-body characteristics are strongly coupled due to the flexible nature of the wing. In order to understand the coupling effect of nonlinear aeroelastic and flight dynamics of the very flexible aircraft, several simulation tools and analysis method have been developed such as ASWING, UM/NAST, NATASHA and NANSI.

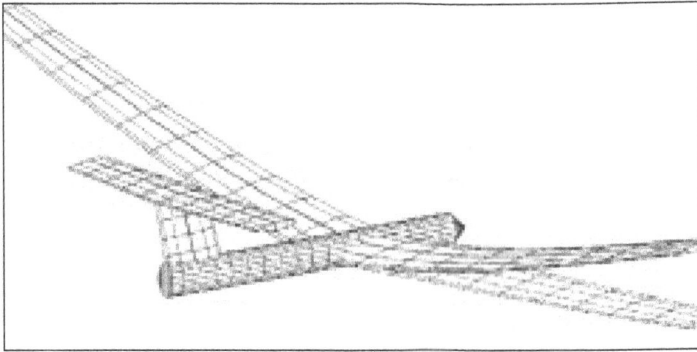

Sketch of T-tail flexible aircraft deformation.

ASWING was initially presented by Drela as an integrated simulation model for flexible aircraft of preliminary aerodynamic, structural, and control-law design. Its structural model is Minguet's nonlinear beam model, and aerodynamic is calculated by vortex lattice method considering Prandtl-Glauert correction for compressibility effect. ASWING aims to conduct rapi modeling and flight simulation of flexible aircraft including nonlinear static, dynamic response with gust fields and eigenmode analysis. Love et al. used ASWING for aero-structural analysis of a sweep flying wing SensorCraft so as to better understand the effect of passive means on body freedom flutter. González et al. modeled the Unmanned Airplane for Ecological Conservation as a flexible-body using the ASWING code and compared it with results from an analytical empirical method and potential flow codes, aiming to evaluate the aerodynamic and static stability of the aircraft. However, the ASWING is limited in capability for nonlinear aeroelastic analysis of joined-wing configurations aircraft.

UM/NAST (The University of Michigan's Nonlinear Aeroelastic Simulation Toolbox) was presented by Cesnik and Brown to study the coupling effects of nonlinear aeroelastic and flight dynamic. They used a computationally effective strain-based structural formulation and the finite-state unsteady subsonic aerodynamic model. The code serves as a plant representation for HALE aircraft control design. It focuses on a reduced number of states to represent the complex nonlinear problem. Then some improvements were suggested by Shearer el al. and Su and Cesnik. The UM/NAST is capable of nonlinear aeroelastic modeling, integral wing actuation for generating maneuver loads, flutter boundary enhancement, gust load alleviation, trajectory control and overall nonlinear vehicle optimization of unconventional configurations. However, the code has not been validated by any flight experimental data. A scaled test HALE aircraft called X-HALE is in the process of development for the purpose of collecting flight test data and subsequently uncovering the strengths and weaknesses of UM/NAST. The UM/NAST is capable of analyzing the nonlinear aeroelastic behavior of both conventional and unconventional configurations aircraft such as flying-wing configuration, and joined-wing configuration, etc.

NATASHA, a computer program named Nonlinear Aeroelastic Trim and Stability of HALE Aircraft, was firstly presented by Patil and Hodges based on geometrically-exact fully intrinsic beam theory and 2-D the finite state induced-flow model of Peters. The initial program could provide trim, payload distribution, stability analysis and dynamic response for conventional and flying-wings configurations with dynamic stall effect unconsidered. Then Patil et al. applied the methodology in the nonlinear gust response in frequency domain and time domain of highly flexible aircraft. Chang and Hodges conducted a simulation of ground vibration test (GVT) environment and presented an analysis and parametric study of the flight dynamics of flexible aircraft. Mardanpour et al. studied the effect of engine placement on aeroelastic trim and stability of flying wing aircraft. Moreover, the capability of

NATASHA was updated by Sotoudeh et al. and applied to statically indeterminate configuration such as joined-wing using incremental discretization method. The results from NATASHA were validated with a range of results from the well-known solutions of beam stability and vibration problems, published experiment data from scaled wind tunnel tests and results from rotorcraft comprehensive analysis system (RCAS). Although the capabilities and validation of NATASHA is still under development, the NATASHA's results are hoped to be used as benchmarks for their own codes.

NANSI is a computational aeroelastic tool presented by Wang et al. for Nonlinear-Aerodynamics/Nonlinear-Structure Interaction. The initial methodology was integrated via tightly coupling a geometrically exact nonlinear intrinsic beam model and the generalized unsteady nonlinear vortex-lattice aerodynamic model, and had the capability of nonlinear time-domain aeroelastic simulation. Then it was enhanced to include the capability of handing effect of gust and stall flow. However, the NANAI is still limited in its capabilities and seems a bit inconvenient for analyzing in frequency-domain.

Additionally, some other simulation frames have been developed to analyze the coupled behavior of flight dynamics and aeroelasticity. Zhao and Ren applied multi-body dynamic approach in flexible aircraft structural modeling and studied the nonlinear aeroelastic characteristic coupled with ONERA aerodynamic model. Zhang and Xiang completed a rigid-flexible coupling simulation frame, considering geometrical nonlinearities, dynamic stall, material anisotropy and rigid body motion with elastic motion of fuselage neglected, via fully intrinsic beam theory and Extended-ONERA aerodynamic model.

Recently, considerable research efforts have been made to know more about the nonlinear aeroelastic phenomena of HALE aircraft. However, the problem of nonlinear aeroelasticity coupled with nonlinear flight dynamics is still not completely followed. Simulation tools have been presented to improve predictions of HALE aircraft response, stability, and overall performance. However, these codes have still been under development and none of them have been completed validated with real flight test data from a HALE aircraft. Mostly, they are validated in a piecemeal fashion against beam models such as a simple cantilevered beam model and wind tunnel data. This is because currently there is scarcely any aircraft flight data available for validation. Consequently, these nonlinear aeroelastic solvers are mainly applied to conceptual design and analysis of HALE aircraft. Among these tools, flexible beam theory and strip theory are the main approaches for aeroelastic modeling for the reason of rapid modeling and solving.

Fight Aircraft

Aeroelastic systems of the fight aircraft are inherently nonlinear due to both aerodynamic and structural nonlinearity. Nonlinear aeroelasticity in structural aspects is mainly caused by control surface freeplay, and store-induced LCO. Due to the complexity of aircraft structure, it is almost impossible to develop a theoretical technique suitable for the nonlinear aeroelastic analysis of full aircraft. Computational aeroelasticity including CFD and CSD, also called fluid-structure interaction (FSI) techniques, is an effective way to solve the aeroelastic problems of full aircraft. A virtual flight test (VFT) technique based on FSI utilizing physics-based modeling and simulation was developed to provide the capability of predicting aeroelastic phenomena on complex full-aircraft models. And it was used to analyze the influence of F-16 store on LCO. Structural nonlinearities at the wing-fold hinge represented by a bilinear spring, and the outboard leading-edge flap hinge represented by a freeplay were investigated. For the bilinear nonlinearity, limit-cycle oscillations can occur with considerable increase in flutter speed above that for nominal hinge stiffness. For the freeplay nonlinearity, it was shown that limit-cycle oscillations are possible within a small range before the critical flutter speed. A medium-fidelity transonic small-disturbance aerodynamic theory was used to investigate the LCO of F-16. Higher-order spectral analysis was performed to identify nonlinear aeroelastic phenomena that are associated with LCO encountered in the F-16 flight test. The results show that nonlinearities associated with LCO are most observable at the forward locations on the wing-tip and underwing launchers. In the vertical direction, the nonlinearity is cubic and leads to the generation of a third harmonic component. In the lateral direction, the nonlinearity is quadratic and leads to the generation of the second and higher-order harmonics. Methods that can identify the onset of nonlinear aeroelastic phenomena, such as LCO have been developed for the flight flutter test data of F/A-18 by Silva and Dunn. Standard correlation and power spectral density techniques, experimentally-identified impulse responses and higher-order spectral techniques are successfully applied to the data.

For the fight aircraft with stealth, there is no Based on the classical flutter solution and an empirical buzz criterion with the Den Hartog equivalent spring equation, flight test techniques for max-freeplay flutter testing were developed for F-22. Then successful flight flutter tests program were flown to meet the prime objectives of flutter testing with maximum freeplay on aileron, rudder, and horizontal while the control surface's hinge moment was at or near zero. The maximum freeplay test results, in spite of the earlier wind tunnel test results, indicated that the design for freeplay for both flutter and buzz may be conservative, and that some relief in freeplay limits may be possible. So flight test is necessary in the design of control surface freeplay. Although there is not many published literature as the military secrecy, experimental test and verification on nonlinear aeroelasticity are necessary for the high performance fight vehicles.

Stall Flutter

Stall flutter is a nonlinear aeroelastic phenomenon with self-excited oscillation of limited amplitude. Those oscillations are called limit cycle oscillations (LCOs). And stall flutter is a LCO occurring as a result of the dynamic stall, which is an aerodynamic nonlinearity. Two dynamic stall regimes are defined: moderate stall and deep stall. Moderate stall is a phenomenon similar to static stall. Deep stall is characterized by the maximum lift and moment values far in excess of their static counterparts. The oscillations of stall flutter associated with moderate stall or deep stall can result in large oscillatory fatigue loads on the wind turbine blade or helicopter rotor blade and thus reduce blade life span as well as harming the safety and the efficiency of the blade operation. Therefore, there had been an interest in the stall flutter suppression for years.

For the past decades, flutter control has been studied by a number of researchers. Yan and Xin studied the oscillatory blowing control of the airfoil flutter by numerical simulation. Nonlinear control methods were reviewed by Kurdila and Akella for high-energy limit cycle oscillations. They concluded that full and partial feedback control or nonlinear model reference adaptive control (MRAC) methodologies were effective to control LCOs. Strganac et al. suggested the adaptive control as the appropriate control strategy for limit cycle oscillations in aeroelastic systems. Those studies mainly emphasized the design of the control algorithms.

Recently, there has been an increasing interest in flutter control via the smart blade, which is the blade equipped with the appropriate and deployed actuator. Various kinds of such actuators have been investigated for flutter suppression. Tang and Dowell completed the LCO suppression of a high-aspect ratio wing by a controllable slender body tip mass distribution. Li and Fleeter adopted piezoelectric actuators for active suppression of nonlinear stall flutter, including limit cycle and chaotic and quasi-periodic separated flow induced airfoil vibration. Sun et al. studied stall flutter and flutter suppression control systems with a novel state-space model description. Liu investigated blade pitch actuation in the aeroservoelastic control of stall-induced flutter of wind turbine blade section. In addition, the trailing edge flap has been widely used as the control surface to reject the unwanted oscillatory loads and control the aeroelastic instability of the blade.

However, the flutter control using the microtab is rarely studied. The microtab, earlier proposed by Yen et al., is simply short flat plates (approximately 1–5% of the airfoil chord in height) attached to the trailing edge perpendicular to the chord line on the pressure or suction side of the airfoil. In contrast to many other actuators for addressing blade loading issues that seem to be complex and energy-consuming, the potential for the microtab to regulate blade loads with high efficiency has been numerically and experimentally studied by researchers. First, the effect of microtab was studied to control small sinusoidal aerodynamic loads on the blade section. Then, the microtab was used to control aerodynamic load of the blade section at low wind speeds associated with operating regions of a typical wind turbine. The microtab controller was also designed to have full controllability on pitch motion. Moreover, the microtab was recently investigated for the control of classic flutter at relatively small angle of attack.

However, the microtab control of the blade section undergoing dynamic stall has not been studied extensively. In most of the previous works with microtab control application, the static aerodynamic force coefficients have been considered as a baseline for the airfoil loading behavior or a small and smooth change in the unsteady aerodynamic loading to control the aeroelastic response of the blade section only in the attached flow. Nevertheless, it has been proved that, with the occurrence of the dynamic stall, the airfoil section suffers a dramatic change in aerodynamic loads in short time intervals, which will worsen the control effect and even cause the instability of the control system. Thus, it is essential to develop new microtab controller to address stall flutter in different dynamic stall regimes.

Aeroelastic Model of Stall Flutter System

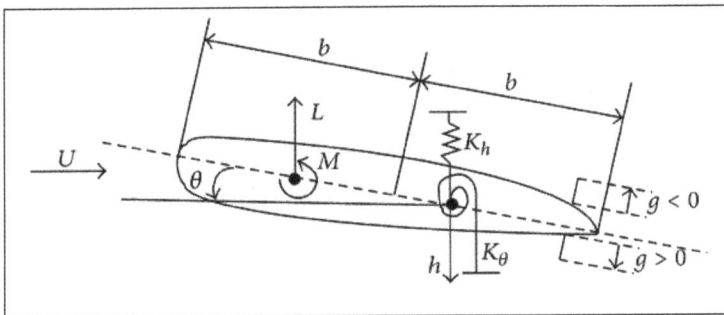

Typical two-dimensional blade section.

In this study, an aeroelastic model of the blade section shown in figure is used for aeroelastic analysis and control design. This is a typical two-dimensional wing section with two degrees of freedom, that is, the plunge motion and the pitch motion. The plunge motion is defined by the plunge displacement h with the direction of positive downwards, and the pitch motion is defined by the pitch angle θ with the direction of positive nose down. The aerodynamic control device, microtabs, is installed close to the trailing edge of the blade. The microtab deployment height g , is regarded as a control input to the system and acts in a similar way to the gurney flap. b is the mid-chord length. K_h and K_θ are plunge stiffness and pitch stiffness, restricted to the elastic axis and C_h and C_θ are the plunge and pitch damping coefficients. The section is exposed to the aerodynamic forces L and aerodynamic moment M at the aerodynamic centre.

Dynamic Stall Model

To include the loading from the unsteady dynamic stall phenomena, the Beddoes-Leishman (B-L) model is adopted to account for the effect of the separated flow at the stall region. Since the basic assumption in B-L model of static lift for trailing edge separation is the lift on a flat plate in a potential Kirchhoff flow, the separation point f^{st} can be defined as follows:

$$f^{st} = \left(2\sqrt{\frac{C_L^{st}(\alpha)}{C_{L,\alpha}(\alpha-\alpha_0)} - 1} \right)^2$$

where C_L^{st} is the static lift coefficient for the airfoil and f^{st} indicates a fully separated flow when it equals zero.

Based on the separation point, the fully separated flow is described as:

$$C_L^{fs} = \frac{C_L^{st} - C_{L,\alpha}(\alpha-\alpha_0)f^{st}}{1-f^{st}}.$$

Note that there will be a problem when $f^{st} = 1$. The value of C_L^{fs} can be approximated as $C_L^{fs}(\alpha)/2$ when f^{st} gets close to one.

The dynamic behavior of the trailing edge separation can be described by the separation related to the pressure distribution over the airfoil section and the separation due to the dynamics of the boundary layer. For the pressure distribution separation, it is assumed that there is a time lag between the unsteady lift coefficient $C_L^{P'}$ and the actual lift coefficient C_L^P The unsteady lift coefficient is modeled as:

$$C_L^{P'} + T_p^{-1}C_L^{P'} = T_p^{-1}C_L^P(t),$$

where T_p is the time constant for the pressure lag and $C_L = 2\pi(\alpha_E - \alpha) + \pi b\theta/U$, α_0 is the angle of attack at zero lift.

The overall dynamics of the flow separation can be revealed by the unsteady separation point, which is caused by the boundary layer dynamics, and involves the effect of the pressure distribution. The equation for the unsteady separation point f^{dy} which makes the separation point lag behind the quasi-steady value f^{ds}, is given as:

$$\dot{f}^{dy} + T_f^{-1}f^{dy} = T_f^{-1}f^{qs}(t),$$

where T_f is the time delay in the boundary layer; the equivalent quasi-steady separation point $f^{qs} = f^{st}(\alpha_f), \alpha_f = C_L^{P'}/C_{L,\alpha} + \alpha_0$ is an equivalent angle of attack that gives the same quasi-steady value of unsteady lift coefficient $C_L^{P'}$.

The unsteady aerodynamic forces of the blade section in stall should involve the unsteady forces induced by both attached flow and the separation flow. The unsteady lift curve can be represented by a liner interpolation between the lift for the fully attached flow and the lift for the fully separated flow as:

$$L_s = \rho U^2 b C_{L0} + L_1 f^{dy} + \rho U^2 b C_L^{fs}(\alpha_E)(1 - f^{dy})$$

where L_s is the unsteady lift including the effect of trailing edge separation, C_{L0} is the static lift coefficients at zero angle of attack, L_1 is the lift for the fully attached flow, and α_E is the effective angle of attack and can be given as:

$$\alpha_E(t) = \left(\frac{b\dot{h}(t)}{U} + \theta(t) + \frac{b\dot{\theta}(t)}{U}\right)(1 - A_1 - A_2) + x_1 + x_2,$$

$$x_1 = b_1 A_1 \frac{1}{bU} \int_0^t \omega_{3/4}(t')U(t')e^{-b_1/b\int_{t'}^t U(\tau)d\tau} dt',$$

$$x_2 = b_2 A_2 \frac{1}{bU} \int_0^t \omega_{3/4}(t')U(t')e^{-b_2/b\int_{t'}^t U(\tau)d\tau} dt',$$

where, x_1 and x_2, yielding the effective downwash, are two state variables of the attached flow; $\omega_{3/4}$ is the downwash at the three-quarter point. $A_1, A_2, b_1,$ and b_1 and b_2 are constant, indicating two time lags in Wagner function of $\phi(s) = 1 - A_1 e^{-b_1 s} - A_2 e^{-b_2 s}$.

The unsteady moment is affected by the trailing edge separation through movement of the pressure center due to the separation. Define the equivalent pressure center as:

$$a^{st} = \frac{C_M^{st} - C_{Mo}}{C_L^{st}},$$

where C_M^{st} and C_L^{st} are the static lift and moment coefficient and C_{M_0} is the moment coefficient at zero lift. Thus, the unsteady moment is given as:

$$M_s = 2\rho U^2 b^2 C_{M0} + M_1$$
$$+ 2\rho U^2 b^2 \left(L_s\left(a^{st}\left(f^{dy}\right) - a^{st}\left(f^{st}(\alpha_E)\right)\right)\right),$$

where the first term is the static moment at zero angle of attack, the second term M_1 is the moment for the fully attached flow, and the last term is the unsteady moment due to the dynamic trailing edge separation.

Assume that the dynamic stall model is linearized with small amplitude vibrations about the equilibrium, given by the steady angle of attack α^0. Then the linearized unsteady lift and moment using Taylor expansion about the steady angle of attack are approximated as:

$$L_s = \rho U^2 b C_L^0 + L_1 f_0 + \rho U^2 b \alpha_E \left(\left. \frac{dC_L^{fs}}{d\alpha} \right|_{\alpha=\alpha^0} (1 - f_0) \right)$$

$$+ \rho U^2 b \left(C_{L,\alpha} \left(\alpha^0 - \alpha_0 \right) - C_L^{fs} \left(\alpha^0 \right) \right) f^{dy},$$

$$M_s = 2\rho U^2 b^2 C_M^0 + M_1 - 2\rho U^2 b^2 \alpha_E \left(C_L^0 \left. \frac{df^{st}}{d\alpha} \right|_{\alpha=\alpha^0} \left. \frac{da^{st}}{df} \right|_{\alpha=\alpha^0} \right)$$

$$+ 2\rho U^2 b^2 C_L^0 \left. \frac{da^{st}}{df} \right|_{\alpha=\alpha^0} f^{dy},$$

where C_L^0 and C_M^0 are the static lift and moment coefficients at the steady angle of attack. f_0 is the static separation point at the steady angle of attack. The details of the formulation of the attached lift L_1 and moment M_1 can be found in the literature.

Governing Equations of the System

With the dynamic stall modeled, the equations of the pitch and plunge motions of the smart blade section undergoing the stall dynamics are as follows:

$$\begin{bmatrix} m & mbx_\theta \\ mbx_\theta & I_\theta \end{bmatrix} \begin{bmatrix} \ddot{h} \\ \ddot{\theta} \end{bmatrix} + \begin{bmatrix} C_h & 0 \\ 0 & C_\theta \end{bmatrix} \begin{bmatrix} \dot{h} \\ \dot{\theta} \end{bmatrix} + \begin{bmatrix} K_h & 0 \\ 0 & K_\theta \end{bmatrix} \begin{bmatrix} h \\ \theta \end{bmatrix}$$

$$= \begin{bmatrix} -L_s \\ M_s \end{bmatrix} + \begin{bmatrix} -L_g \\ M_g \end{bmatrix},$$

where, m is the mass of the airfoil section, x_θ is the dimensionless distance between the elastic axis and the mass center, and I_θ is the inertia moment about the elastic axis.

Define $z = [h \ \theta]^T$, $F = [-L_s \ M_s]^T$, and $F_g = [-L_g \ M_g]^T$; the previous equation of the structural model can be rewritten in the form as:

$$M_0 \ddot{z} + C_0 \dot{z} + K_0 z = F + F_g,$$

where, F is the unsteady aerodynamic forces, F_g is the aerodynamic forces due to control surface, $L_g = \rho U^2 b C_{L,g} g$ and $M_g = 2\rho U^2 b^2 C_{m,g} g$ represent the change in the airfoil lift and pitching moment due to the microtab deployment, and $C_{L,g}$ and $C_{M,g}$ re the corresponding aerodynamic coefficients. These coefficients were calculated by numerical simulation using the commercial CFD code, FLUENT.

Defining X_a as the vector of x_1 and x_2 and X_s as the vector of $C_L^{P'}$ and f^{dy},

$$L_s = \rho U^2 b C_L^0 + L_1 f_0 + \rho U^2 b \alpha_E \left(\frac{dC_L^{fs}}{d\alpha} \bigg|_{\alpha=\alpha^0} (1-f_0) \right)$$

$$+ \rho U^2 b \left(C_{L,\alpha} \left(\alpha^0 - \alpha_0 \right) - C_L^{fs} \left(\alpha^0 \right) \right) f^{dy},$$

$$M_s = 2\rho U^2 b^2 C_M^0 + M_1 - 2\rho U^2 b^2 \alpha_E \left(C_L^0 \frac{df^{st}}{d\alpha} \bigg|_{\alpha=\alpha^0} \frac{da^{st}}{df} \bigg|_{\alpha=\alpha^0} \right)$$

$$+ 2\rho U^2 b^2 C_L^0 \frac{da^{st}}{df} \bigg|_{\alpha=\alpha^0} f^{dy},$$

can be rewritten in the matrix form as:

$$F = \begin{bmatrix} -L_s \\ M_s \end{bmatrix} = D_1 z + D_2 \dot{z} + C_1 X_a + C_2 X_s + C_0,$$

where,

$$D_1 = \begin{bmatrix} 0 - \rho U^2 b \left(\bar{a} C_{L,\alpha} f_0 + \bar{a} C_{L,\alpha^0}^{fs} (1-f_0) \right) \\ 0 \quad 2\rho U^2 b^2 \left(\bar{a} C_{M,\alpha^0}^{st} - \bar{a} C_L^0 f_{\alpha^0}^{st} a_{\alpha^0}^{st} \right) \end{bmatrix},$$

$$D_2 = \begin{bmatrix} -\rho U \bar{a} b^2 \left(C_{L,\alpha} f_0 + C_{L,\alpha^0}^{fs} (1-f_0) \right) - \rho U \bar{a} b^2 \left(C_{L,\alpha} f_0 + C_{L,\alpha^0}^{fs} (1-f_0) + \frac{\pi}{a} \right) \\ 2\rho U \bar{a} b^3 \left(C_{M,\alpha^0}^{st} - C_L^0 f_{\alpha^0}^{st} a_{\alpha^0}^{st} \right) \quad 2\rho U \bar{a} b^3 \left(C_{M,\alpha^0}^{st} - C_L^0 f_{\alpha^0}^{st} a_{\alpha^0}^{st} - \frac{\pi}{2\bar{a}} \right) \end{bmatrix}$$

$$C_1 = \begin{bmatrix} -\rho U^2 b \left(C_{L,\alpha} f_0 + C_{L,\alpha^0}^{fs} (1-f_0) \right) - \rho U^2 b \left(C_{L,\alpha} f_0 + C_{L,\alpha^0}^{fs} (1-f_0) \right) \\ 2\rho U^2 b^2 \left(C_{M,\alpha^0}^{st} - C_L^0 f_{\alpha^0}^{st} a_{\alpha^0}^{st} \right) \quad 2\rho U^2 b^2 \left(C_{M,\alpha^0}^{st} - C_L^0 f_{\alpha^0}^{st} a_{\alpha^0}^{st} \right) \end{bmatrix}$$

$$C_2 = \begin{bmatrix} 0 & -\rho U^2 b \left(C_{L,\alpha} \left(\alpha^0 - \alpha_0 \right) - C_L^{fs} \left(\alpha^0 \right) \right) \\ 0 & 2\rho U^2 b^2 C_L^0 a_{\alpha^0}^{st} \end{bmatrix}$$

$$C_0 = \begin{bmatrix} -\rho U^2 b C_L^0 \\ 2\rho U^2 b^2 C_M^0 \end{bmatrix}$$

$$\bar{a} = 1 - A_1 - A_2$$

$$C_{L,\alpha^0}^{fs} = \left. \frac{dC_L^{fs}}{d\alpha} \right|_{\alpha=\alpha^0}$$

$$C_{M,\alpha^0}^{st} = \left. \frac{dC_M^{st}}{d\alpha} \right|_{\alpha=\alpha^0}$$

$$f_{\alpha^0}^{st} = \left. \frac{df^{st}}{d\alpha} \right|_{\alpha=\alpha^0}$$

$$a_{\alpha^0}^{st} = \left. \frac{da^{st}}{df} \right|_{\alpha=\alpha^0}.$$

With the definition of X_a and X_s,

$$C_L^{P'} + T_p^{-1} C_L^{p'} = T_p^{-1} C_L^p(t), \dot{f}^{dy} + T_f^{-1} f^{dy} = T_f^{-1} f^{qs}(t), \alpha_E(t)$$

$$= \left(\frac{b\dot{h}(t)}{U} + \theta(t) + \frac{b\dot{\theta}(t)}{U} \right)(1 - A_1 - A_2) + x_1 + x_2 \text{ and}$$

$$x_1 = b_1 A_1 \frac{1}{bU} \int_0^t \omega_{3/4}(t') U(t') e^{-b_1/b \int_{t'}^t U(\tau)d\tau} dt',$$

$$x_2 = b_2 A_2 \frac{1}{bU} \int_0^t \omega_{3/4}(t') U(t') e^{-b_2/b \int_{t'}^t U(\tau)d\tau} dt',$$

of the aerodynamic stall model can be rewritten in the state-space forms as:

$$\dot{X}_s = A_3 X_a + A_4 X_s + B_3 z + B_4 \dot{z},$$
$$\dot{X}_a = A_1 X_a + A_2 X_s + B_1 z + B_2 \dot{z},$$

where,

$$A_1 = \begin{bmatrix} -T_1^{-1} & 0 \\ 0 & -T_2^{-1} \end{bmatrix}$$

$$A_2 = \begin{bmatrix} 0 & 0 \\ 0 & 0 \end{bmatrix}$$

$$A_3 = \begin{bmatrix} T_p^{-1} C_{L,\alpha} & T_p^{-1} C_{L,\alpha} \\ 0 & 0 \end{bmatrix}$$

$$A_4 = \begin{bmatrix} -T_p^{-1} & 0 \\ \dfrac{T_f^{-1}}{C_{L,\alpha}} f_{\alpha^0}^{st} & -T_f^{-1} \end{bmatrix}$$

$$B_1 = \begin{bmatrix} 0 & T_1^{-1} A_1 \\ 0 & T_2^{-1} A_2 \end{bmatrix}$$

$$B_2 = \begin{bmatrix} \dfrac{T_1^{-1} A_1 b}{U} & \dfrac{T_1^{-1} A_1 b}{U} \\ \dfrac{T_2^{-1} A_2 b}{U} & \dfrac{T_2^{-1} A_2 b}{U} \end{bmatrix},$$

$$B_3 = \begin{bmatrix} 0 & \bar{a} T_p^{-1} C_{L,\alpha} \\ 0 & 0 \end{bmatrix}$$

$$B_4 = \begin{bmatrix} \dfrac{\bar{a} b T_p^{-1} C_{L,\alpha}}{U} & \dfrac{\bar{a} b T_p^{-1} C_{L,\alpha}}{U} + \dfrac{\pi T_p^{-1} b}{U} \\ 0 & 0 \end{bmatrix}.$$

The aeroservoelastic model is a combination of the structural model in $M_0 \ddot{z} + C_0 \dot{z} + K_0 z = F + F_g$, and the aerodynamic model in. Defining the state variables of the system as $q = \begin{bmatrix} z & \dot{z} & X_a & X_s \end{bmatrix}^T$, the governing equations of the aeroservoelastic model can be written in state-space form as:

$$\begin{bmatrix} \dot{z} \\ \ddot{z} \\ \dot{X}_a \\ \dot{X}_s \end{bmatrix} = \underbrace{\begin{bmatrix} 0_{2\times2} & I_{2\times2} & 0_{2\times2} & 0_{2\times2} \\ -M_0^{-1}(K_0 - D_1) & -M_0^{-1}(C_0 - D_2) & M_0^{-1} C_1 & M_0^{-1} C_2 \\ B_1 & B_2 & A_1 & A_2 \\ B_3 & B_4 & A_3 & A_4 \end{bmatrix}}_{A} \begin{bmatrix} z \\ \dot{z} \\ X_a \\ X_s \end{bmatrix} + \underbrace{\begin{bmatrix} 0_{2\times1} \\ G \\ 0_{2\times1} \\ 0_{2\times1} \end{bmatrix}}_{B} g,$$

where,

$$G = \begin{bmatrix} \dfrac{-\rho U^2 b I_\theta C_{L,g} - 2\rho U^2 b^3 m x_\theta C_{M,g}}{m I_\theta - m^2 b^2 x_\theta^2} \\ \dfrac{\rho U^2 b^2 m x_\theta C_{L,g} + 2\rho U^2 b^2 m C_{M,g}}{m I_\theta - m^2 b^2 x_\theta^2} \end{bmatrix}.$$

$$\begin{bmatrix} \dot{z} \\ \ddot{z} \\ \dot{X}_a \\ \dot{X}_s \end{bmatrix} = \underbrace{\begin{bmatrix} 0_{2\times2} & I_{2\times2} & 0_{2\times2} & 0_{2\times2} \\ -M_0^{-1}(K_0 - D_1) & -M_0^{-1}(C_0 - D_2) & M_0^{-1}C_1 & M_0^{-1}C_2 \\ B_1 & B_2 & A_1 & A_2 \\ B_3 & B_4 & A_3 & A_4 \end{bmatrix}}_{A} \begin{bmatrix} z \\ \dot{z} \\ X_a \\ X_s \end{bmatrix} + \underbrace{\begin{bmatrix} 0_{2\times1} \\ G \\ 0_{2\times1} \\ 0_{2\times1} \end{bmatrix}}_{B} g,$$

is a linearized equation, with the state matrix, containing the corresponding matrices from the aerodynamic stall model and the structural model.

Stall Conditions and Stall Flutter Analysis

The effect of different dynamic stall conditions and the consequent aeroelastic behavior of the section undergoing stall flutter. Naughton found that the mean angle of attack is one of the important parameters that affect the behavior of the airfoil in stall significantly. Clearly, the extent of the dynamic stall phenomenon can also affect the aeroelastic behavior of the airfoil. Thus, the study focuses on two different cases of dynamic stall regime. The first case is the moderate stall condition, which holds the mean angle of attack right before the static stall angle $(\alpha_s = 12°)$. The second case is the deep stall condition, which has the mean angle of attack beyond the static stall angle. In deep dynamic stall condition, large levels of hysteresis exist in the lift and moment coefficient curves, as opposed to the moderate dynamic stall case where small hysteresis in aerodynamic coefficients is seen. In order to investigate aeroelastic stability, the following model parameters are used:

$$m = 50 \text{ kg}, \ K_h = 2884 \ N/m, \ C_h = 7.54 \ kg/s, \ K_\theta = 700,$$

$$C_\theta = 0.23 \ kgm^2/s, \ I_\theta = 0.4 \ m^2kg, \ \rho = 1.2 \ kg/m^3, \text{and } b = 0.5m.$$

Moderate Dynamic Stall

Before investigating the aeroelastic response of the airfoil under moderate stall condition, the moderate stall behavior of the blade section needs to be well predicted by the simulation model. The moderate stall aerodynamic curves are simulated based on static experimental data.

Figure shows hysteresis in the dynamic lift and moment curves. The dynamic lift loop has small overshoot based on the static curves, the dynamic moment loop is well surrounding the static curves, and the dynamic curves follow the trend of the static curves as indicated in. For the dynamic lift curve, it is clear that the moderate stall is indicated by a gentle drop between max. point and reattached point and a small time lag between static stall point and max. point. For the dynamic moment curve, there is also a relatively smooth drop between max. point and reattached point. The max. point before the static point is caused by the dynamic moment shape that follows the trend of static curve before the static stall point.

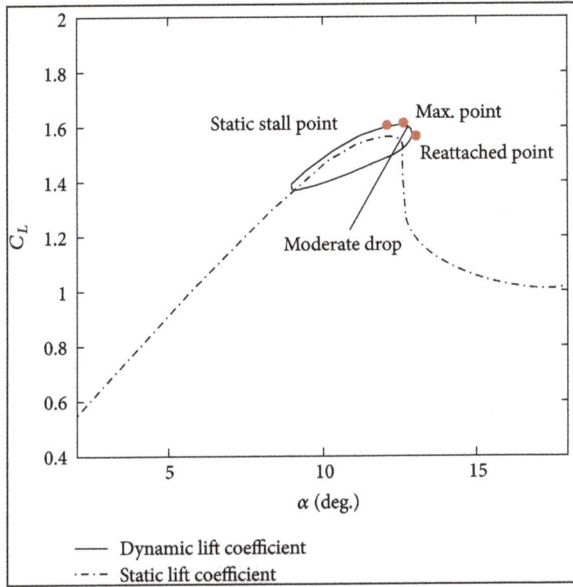

Moderate dynamic stall simulated curve and static experimental
data: Lift force coefficient.

The pitch and plunge responses at moderate stall flutter are shown in figure, where the responses tend to damp at first 30 seconds, and they are stabilized after 40 seconds with asymmetric LCOs (e.g., plunge displacement LCO with magnitude from 0 m to −0.5 m and pitch angle LOC with magnitude from −5 deg. to −10 deg.). This phenomenon is convinced by the experimental results in, which indicated the occurrence of only asymmetric LCOs with either positive or negative pitching centering.

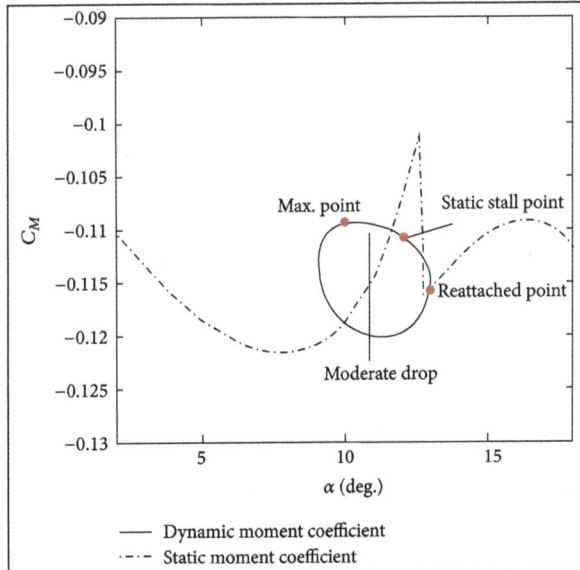

Moderate dynamic stall simulated curve and static experimental
data: moment coefficient.

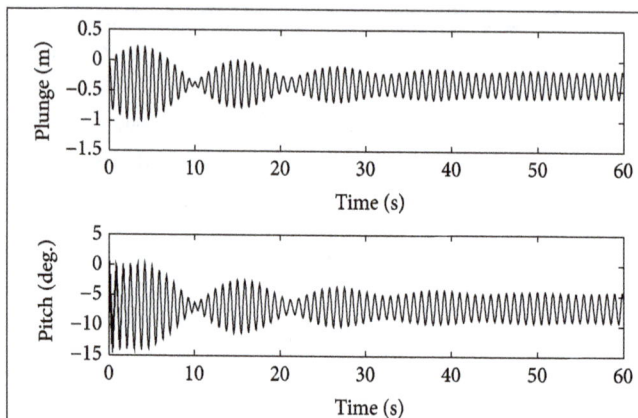

Pitch and plunge responses at moderate stall flutter.

Deep Dynamic Stall

The deep stall aerodynamic curves are also simulated based on static experimental data. Figure shows the deep stall motions of the simulated unsteady lift and moment coefficients at angle of attack $\alpha = 13.5° + 3.5° \sin(10t)$. The deep stall curves are also well predicted by the proposed model to capture the trend of static curves in the neighborhood as expected. In contrast with the moderate stall in figure, the characteristics of deep stall dynamic are revealed by a relatively sharp drop between max. point and reattached point due to the strong separation of the flow and more drastic hysteresis with significant difference between up-stroke and down-stroke phase of dynamic curves. In addition, the maximum lift and moment point are far beyond the static point with a large time lag.

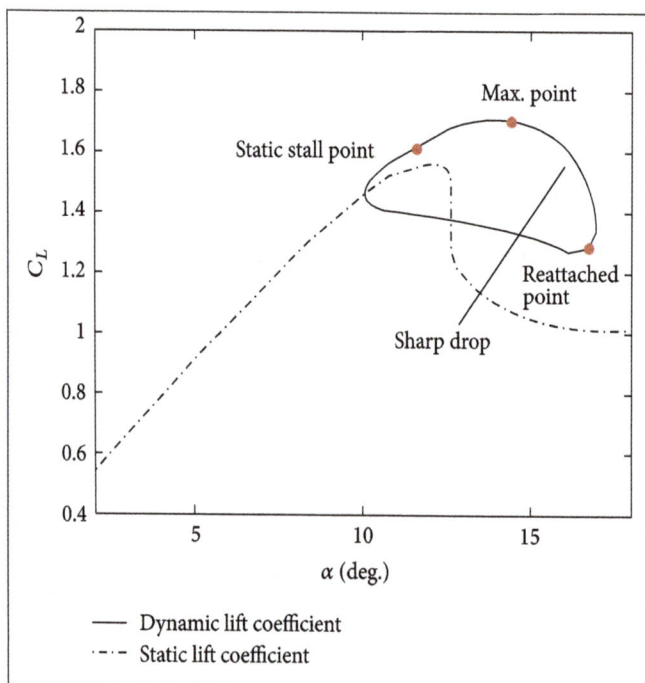

The stall phenomena in figures can also be confirmed by the model validation results in, where the lift loop near max. point has a very different shape owing to the separation effects, and the lift curve has expected overshoot, while moment curves revolve around the static curves.

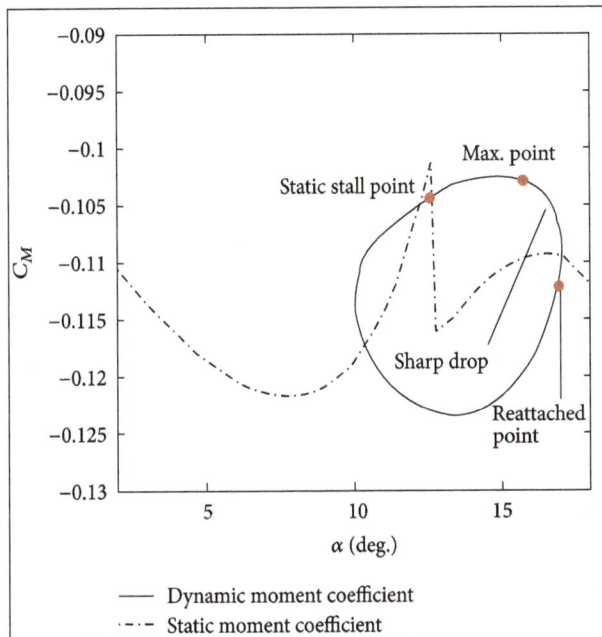

Deep dynamic stall simulated curve and static experimental data. (a) Lift force coefficient and (b) moment coefficient.

Figure shows the pitch and plunge responses at deep stall flutter, where the asymmetric LCOs are also observed and their magnitudes are invariant after 25 seconds, resulting in magnitude of the plunge from 0 m to −1 m and magnitude of the pitch from −5 deg. to −15 deg. Compared with the moderate stall flutter, the deep stall can further decrease the damping ratio of the pitch and plunge modes, leading to relatively larger LCOs in a shorter time.

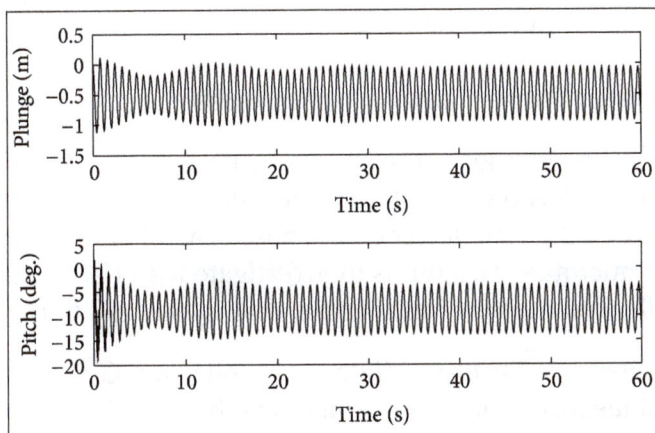

Pitch and plunge responses at deep stall flutter.

The $\theta - \dot{\theta}$ phase plane is illustrated in figure for both moderate stall flutter case and deep stall flutter case. Two asymmetric limit cycle oscillations are observed. It is revealed that the asymmetric LCOs grow in size, and pitching center moves towards the left half of the plane as the section goes through more significant stall hysteresis, which can bring severe oscillations. This phase analysis result is consistent with the results of time-domain responses in figures.

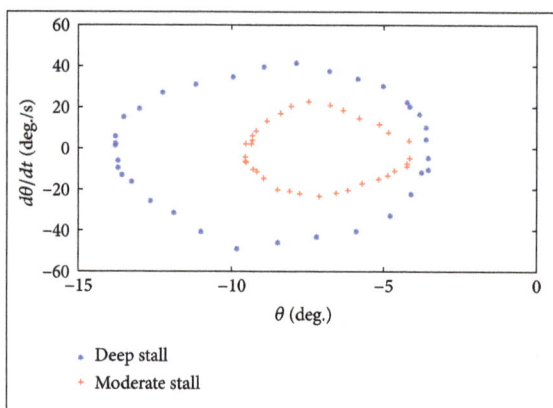

Phase plane of θ and $\dot{\theta}$ of the system at moderate/deep stall flutter.

Stall Flutter Control

Since the stall flutter occurs with the unsteady time-varying aerodynamic forces, which is induced by the time-varying change of angle of attack $\alpha^0(t)$, the aeroservoelastic system can be rewritten as:

$$\dot{X} = A(t)X + Bu$$
$$y = CX,$$

where $A(t)$ is the state matrix, which holds the coefficients in (17), u is the control input, that is, the microtab deployment of the actuator, and the output sensor signal y can be determined by the output matrix C.

The Test of Former Aeroelastic Controllers

Before the study of the control algorithm for stall flutter control, two former controllers that have been developed by the author in classic flutter control study and pitch control study are first tested on the stall flutter case. Those controllers are also developed with the actuator of the microtab. The aim is to investigate the effectiveness of two former aeroelastic controllers on the new control case, that is, the stall flutter control.

- Controller One: It is the classic flutter controller $u = G_e y, y = \dot{\theta}$. This controller is an output feedback controller, which was shown to be effective in classic flutter suppression of the airfoil section.

- Controller Two: It is the adaptive pitch controller.

$$u = G_e y \left(y = \theta + 0.1\dot{\theta}, \dot{G}_e = -\gamma_e y^2 \right).$$

This adaptive control was proved to have full controllability of pitch mode and partial controllability of plunge mode.

Those two aeroelastic controllers are adopted to control the stall flutter of the blade section separately. The control results are shown in figure for moderate stall case and in figure for deep stall case. It can be found that the stall flutter oscillations are not effectively suppressed via the adaptive pitch controller in 5 seconds even though there is tendency of decaying for both stall cases. At the same time, even though the classic flutter controller works for the pitch mode and performs better than the adaptive pitch controller on the plunge mode, the oscillations of the plunge displacement cannot be totally suppressed in 5 seconds as they do for classic flutter case. The better control performance of pitching than plunging is due to control design with system output of pitching signal, as indicated in controller one and controller two. Therefore, those former flutter controllers are not good enough to eliminate the stall flutter effectively in a short time. Stall flutter controller needs to be designed with the investigation of new sensor signal and new control algorithm.

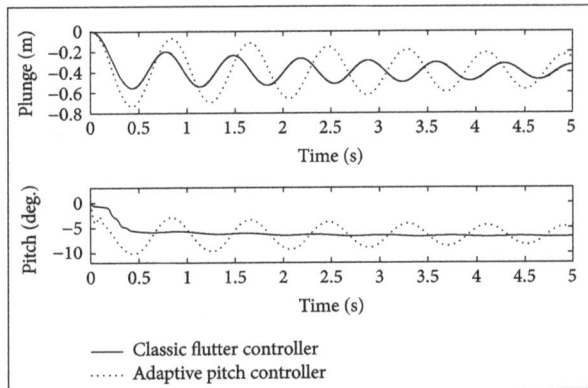

System responses with classic flutter controller and adaptive pitch controller at moderate stall flutter case.

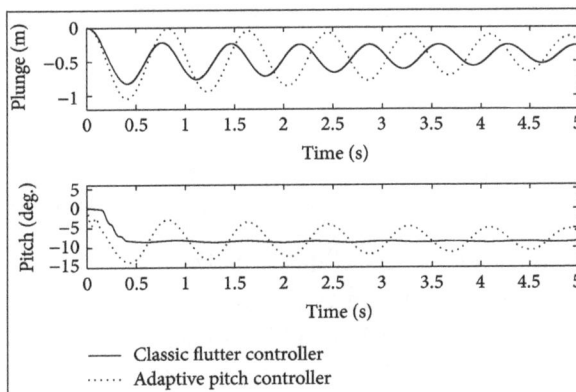

System responses with classic flutter controller and adaptive pitch controller at deep flutter case.

The Design of Stall Flutter Controller

In order to develop the effective stall flutter controller for both stall cases, there are several aspects that need to be investigated: first, the effective sensor signals (or the control input signals for the controller) that can control the aeroelastic behavior of the blade section undergoing dynamic stall significantly, second, the controllability of the microtab (the control capability of the actuator), and third, the powerful control algorithm that can suppress the asymmetric LCOs of the stall flutter in a short time.

Sensor Signal Effect on Stall Flutter Control

To study the sensor signal effect on the control of stall flutter, the eigenvalues of the closed-loop system with a static output feedback of the sensor signal are analyzed. The state matrix of the closed-loop system is taken as $A_c = A + BG_eC, A = A(t)\big|_{\alpha^0(t)=11}$, for moderate stall case and $A = A(t)\big|_{\alpha^0(t)=13.5}$ for deep stall case; $y = Cq$ is the sensor signal. Note that the controllability matrix can also be computed to indicate the sensor signal effect, but here the eigenvalue analysis makes a more vivid explanation. The LCOs of the stall flutter indicate the imaginary stability of open-loop system.

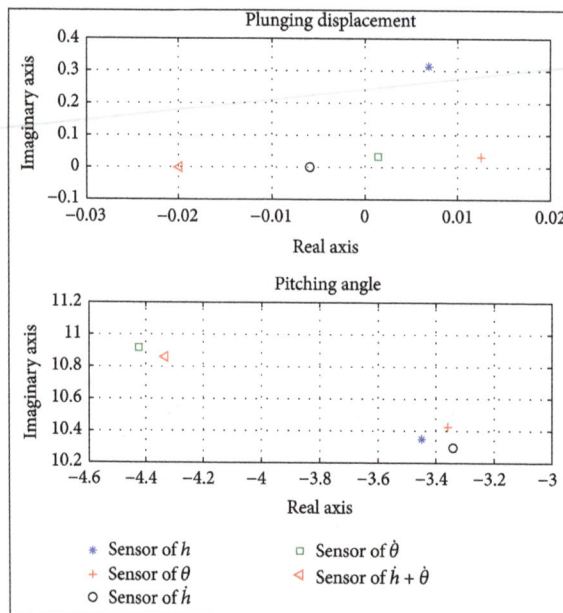

Closed-loop eigenvalues of the system with different sensor signals for moderate stall case.

For moderate stall flutter case, figure illustrates the eigenvalues of plunge and pitch modes in closed-loop systems with different sensor signals. It is revealed that the sensor of h, the sensor of θ, and the sensor of $\dot{\theta}$ all lose the controllability of the plunge mode in stall flutter, even though they can have the full controllability of the pitch mode. Nevertheless, both the sensor of $\dot{h}$ and the sensor of $\dot{h}+\dot{\theta}$ can suppress the LCOs with no imaginary parts in plunge mode and all negative values in pitch mode.

Among those two sensor signals, the sensor of $\dot{h} + \dot{\theta}$ is more influential in control of moderate stall flutter. Note that, for consideration of the sensor number limitation or sensor cost in practice, the single sensor signal of h can be chosen for implementation. For deep stall flutter case, the eigenvalues results shown in figure are similar with those for moderate stall flutter case with little less negative values in pitch mode.

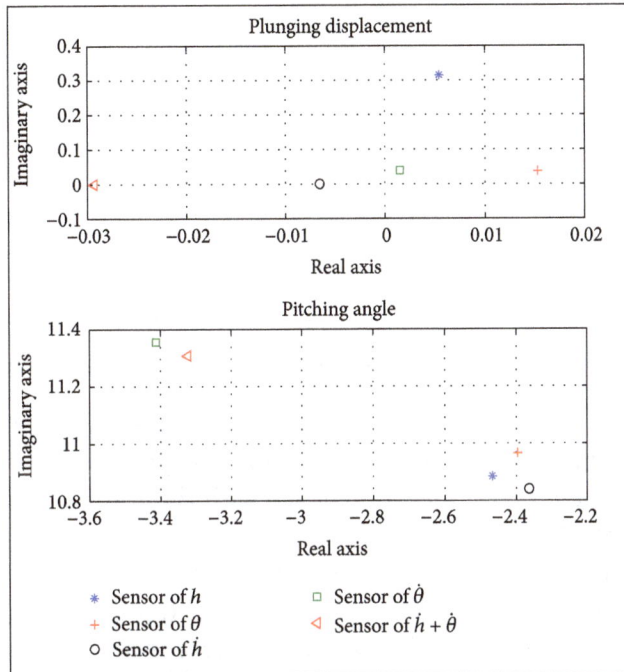

Closed-loop eigenvalues of the system with different sensor signals for deep stall case.

Microtab Control Analysis on Stall Flutter

Since the potential sensor signals are studied for both stall flutter cases, the limitation of the microtab deployment also needs to be considered for practice. The airfoil type is DU97W300-10, and the location of the microtab is at 100% chord. The deployment of the microtab is saturated to 1%–5% of chord. The aerodynamic coefficients are adopted with respect to the deployment of the microtab. Four different limitations of the micro-tab deployment are tested to verify the control capability of the microtab on LCOs of moderate/deep stall flutter.

The damping ratios of plunge and pitch modes in the feedback system with microtab limitations (1% of c, 2% of c, 4% of c, and 5% of c) are illustrated in figure. It is clear that both modes have increasing damping ratios (a sharp increase in plunge damp-ing ratio and a moderate increase in pitch damping ratio) with the growing microtab deployments, indicating that the LCOs in both stall flutter cases can be successfully suppressed within the limitations of the microtab. And the microtab has more control ability on plunge mode than pitch mode.

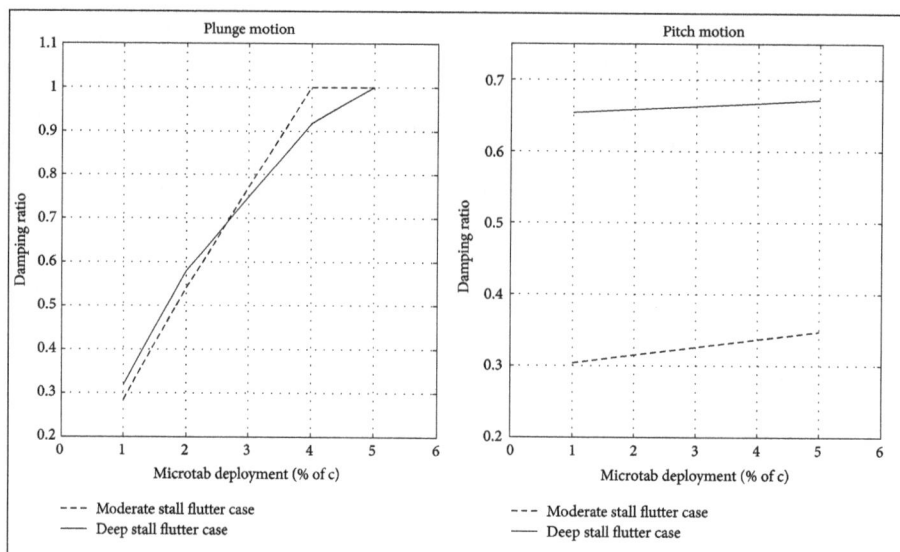

Damping ratios versus microtab limitation for moderate/deep stall flutter cases.

For the control of plunge motion, small microtab limitation (e.g., 1-2% of c) leads to the decaying LCOs with small damping ratios less than one, and full microtab deployment (5% of c) brings completely damped responses with damping ratio of 1 for both stall flutter cases. However, the microtab deployment of 4% of c has a weaker impact on deep stall flutter with less damping ratio than that of moderate stall flutter case, revealing that it is a little harder to suppress the deep stall flutter compared to moderate stall flutter. For the control of pitch motion, although there is a small increase in damping ratio, the microtab seems to have similar effect on the increase of the damping ratio for both stall flutter cases. Thus, microtab is proved as the potential and effective actuator for the stall flutter control, with powerful deployment of 4% of c for moderate stall case and 5% of c deployment for deep stall case.

Stall Flutter Controller

The control objective is to minimize the asymmetric LCOs in the stall flutter. Since the former aeroelastic controllers do not have capability in stall flutter control, the controller needs to be improved with new control strategies or new effective sensor signal for the stall flutter application. Two different stall flutter controllers are developed based on different control algorithms and the new potential sensor signal for stall flutter control.

Simple Stall Flutter Controller: First, the simple output feedback control strategy used for classic flutter control is adopted to design the stall flutter controller along with the new potential sensor signal. Then, the simple stall flutter controller is given as:

$$u = G_e y,$$
$$y = \dot{h} + \dot{\theta}.$$

Note that, on this condition, the stall flutter control system is taken as a single input and single output system.

Adaptive Stall Flutter Controller: To improve the control capability of each vibration mode, the stall flutter control system is treated as a single-input two-output system, and the adaptive control is used as the control algorithm. The block diagram of the adaptive stall flutter control strategy is illustrated in figure. Here, the output of the system is defined as the vector $y = \begin{bmatrix} \dot{h} & \dot{\theta} \end{bmatrix}^T$. Since the system is time-varying, the control gain should also be designed to accommodate with the time-varying change of system parameters.

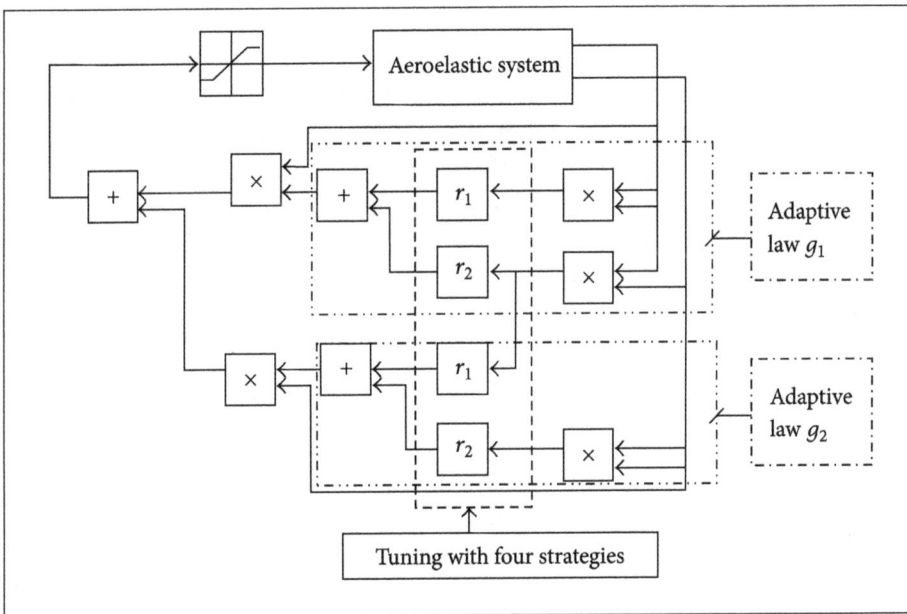

Block diagram of the adaptive stall flutter control strategy.

The adaptive stall flutter controller is presented as:

$$u = G_e(t)y,$$
$$G_e(t) = \begin{bmatrix} g_1(t) & g_2(t) \end{bmatrix},$$

where $G_e(t)$ is the time-varying adaptive control gain, which can be updated by the adaptive law as:

$$g_1(t) = r_1 \dot{h}^2 + r_2 \dot{h}\dot{\theta},$$
$$g_2(t) = r_1 \dot{h}\dot{\theta} + r_2 \dot{\theta}^2,$$

where, r_1 and r_2 are scalar. The adaptive law can regulate the change of adaptive control gain to make plunge displacement h and pitch angle θ asymptotically stable.

Since the value of r_1 and r_2 can impact the performance of the adaptive law, four strategies of r_1 and r_2 are studied for the stall flutter control: case 1, $r_1 = -1$ and $r_2 = -1$, case 2, $r_1 = -1$ and $r_2 = -0.1$, case 3, $r_1 = -0.1$ and $r_2 = -1$, and case 4, $r_1 = -0.5$ and $r_2 = -0.5$. First, the control responses of adaptive stall flutter controllers with case 1, case 2, and case 3 are compared in figure. It can be revealed that the controller with all control strategy cases can suppress the LCOs of plunge and pitch modes in 5 seconds as expected. However, the control strategy of case 2 leads to high-frequency pitch oscillations in first 2 seconds for moderate stall case and even as long as 4 seconds for deep stall case because case 2 lacks control power in pitch mode with $|r_1| \ll |r_1|$. At the same time, case 3 performs well in pitch oscillation suppression, while it performs worse in plunge mode among all three cases because of $|r_1| \ll |r_1|$. Only case 1 which has equivalent weigh on both modes $(|r_1| \ll |r_2|)$ can suppress all LCOs effectively. The unsatisfying control results of case 2 and case 3 are due to the fact that there is not very strong coupling between pitching and plunging under dynamic stall condition. Hence, the individual control of either plunging or pitching cannot suppress them all effectively at the same time. In order to further investigate the effect of r_1 and r_2, the controller with case 4 is compared with that with case 1 in figure. It is shown that the decrease of value of $|r_1|, |r_2|$ brings the similar overshoot of the control response but with a longer settling time, especially for the plunge control. All in all, the adaptive stall flutter control with control strategy of case 1 is the most potential one for LCOs suppression for both stall flutter cases.

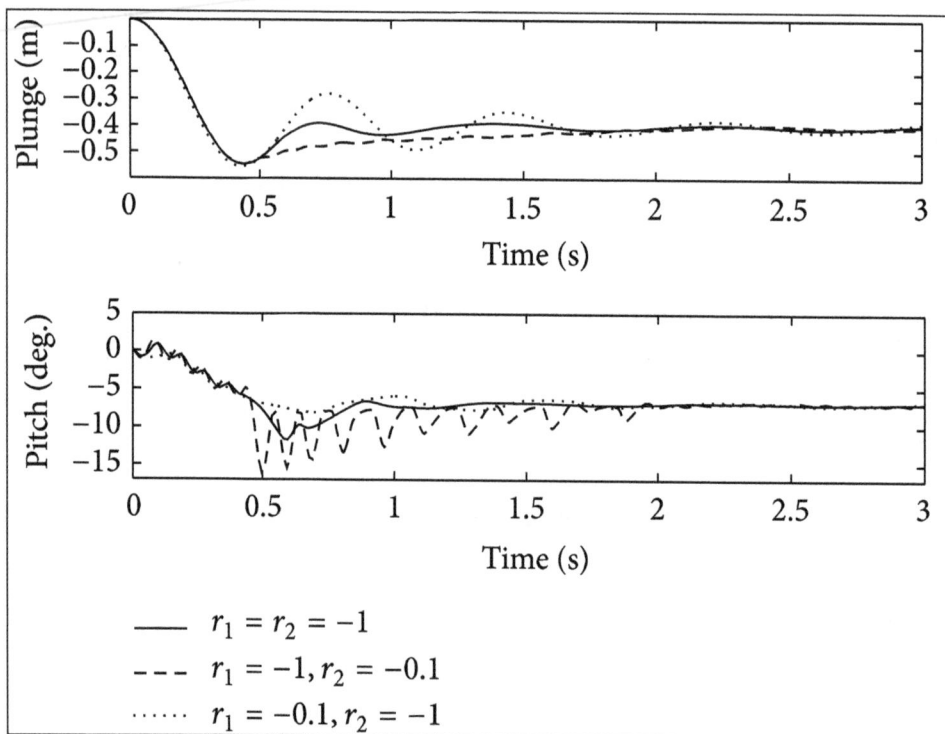

Adaptive stall flutter control responses with different
control strategies for moderate stall flutter case.

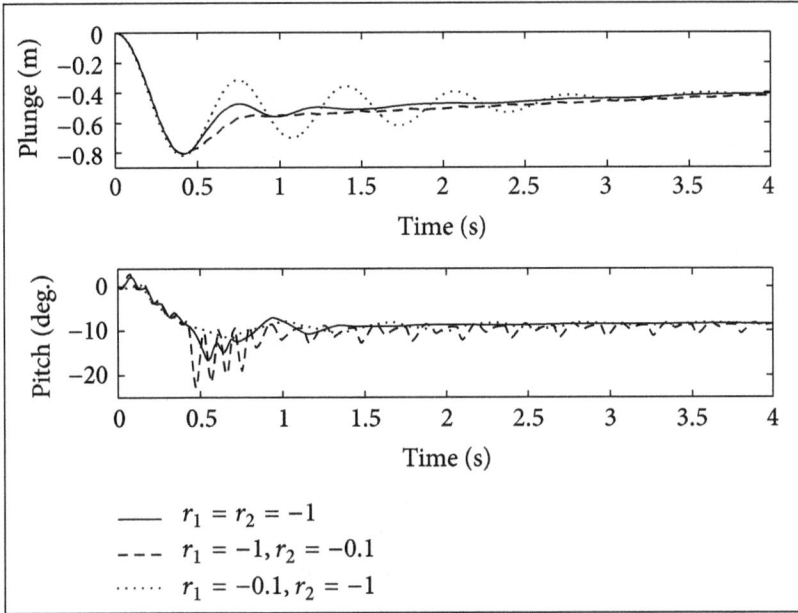

Deep Stall Flutter Case.

Control Results

To show the superiority of the proposed adaptive stall flutter controller, it is compared with the simple stall flutter controller. The results of control performance with those controllers are shown in figure. In addition, the effects of adaptive stall flutter control on moderate stall flutter and deep stall flutter are illustrated in figure.

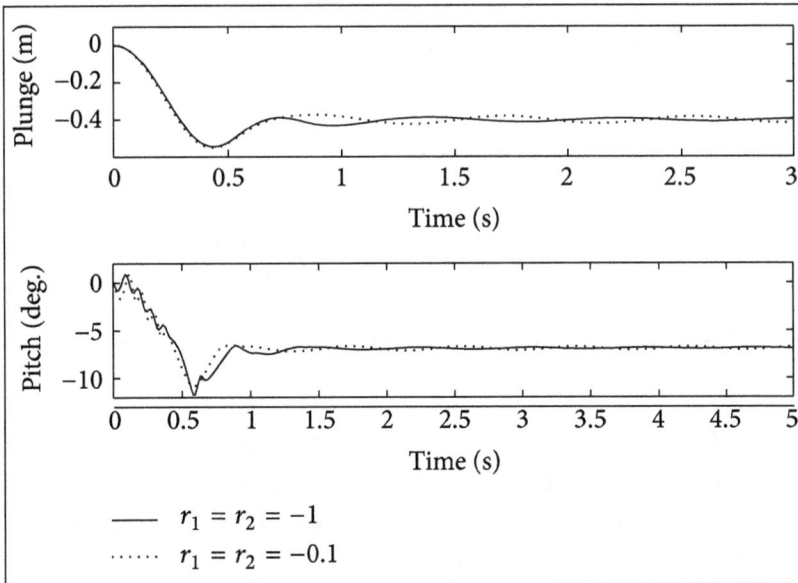

Response of the closed-loop system with simple stall flutter controller and adaptive stall flutter controller for moderate stall flutter case.

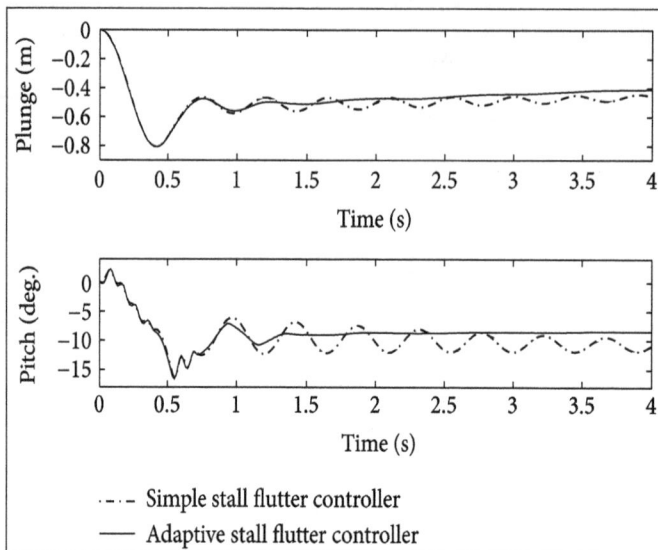

(b) deep stall flutter case.

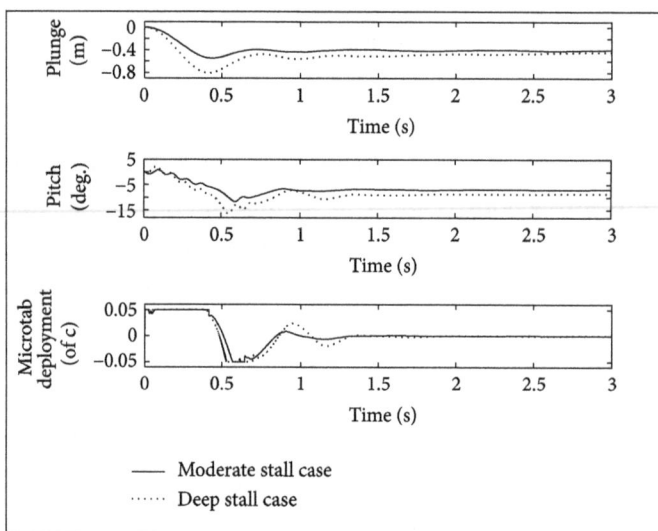

Control results of the closed-loop system
for moderate/deep stall flutter cases.

Simple Stall Flutter Controller versus Adaptive Stall Flutter Controller

Figure shows the control results of the simple flutter controller and the adaptive stall flutter controller. With respect to the classic flutter controller, the simple flutter controller has improved control performance on plunge mode with the new sensor signal. Specifically, the plunge displacement has a quick decay of the oscillations at the first several simulation intervals with a relatively short settling time and small magnitude of the oscillations, which confirms the effectiveness of the new sensor signal in the stall flutter control. In other words, the sensor signal has a significant impact on the improvement of the control effect on different flutter cases even with the same control algorithm.

Although the simple flutter controller can effectively suppress the LCOs and stabilize the plunged mode in about 4 seconds and the pitch mode at the same time, the dynamic performance is not very satisfying with lots of small oscillations around the steady state after 1 second, which indicate that the utilization of only new sensor signal is not powerful enough to eliminate all vibrations in a short time for both modes. In contrast, the adaptive stall flutter controller shows the superiority with the new control algorithm and the new sensor signal. As indicated in figure, for moderate stall case, the proposed adaptive stall flutter controller greatly improves the control performance of pitch and plunge modes. Both vibration responses can converge to the steady state in 2 seconds, and there are no significant oscillations after 1.5 seconds to show the great dynamic performance. However, the static stall flutter controller is not capable of suppressing all oscillations in a short time even though it can stabilize the system with the decaying responses for all stall cases. For deep stall flutter with stronger dynamic stall, simple stall flutter controller performs even worse with larger oscillations in vibration modes, while the adaptive stall flutter control can still suppress the stall flutter quickly with good dynamic performance. Thus, the adaptive stall flutter controller outperforms the static stall flutter controller.

For the physical meaning, first, the poor control performance of the classic flutter controller on pitching control is due to the weak coupling between pitching and plunging on stall flutter cases. Then, this weak coupling is strengthened at specific operation point by the simple stall flutter controller with new sensor signal, which is the most potential sensor signal in stall flutter control. However, as the parameters of the system are varying from one operation point to another, the coupling dynamic is also changing from time to time. On this condition, the simple stall flutter controller cannot adapt to the coupling variance with the constant control gain. Nevertheless, the proposed adaptive stall flutter controller can adapt to the varying coupling conditions at different operation points and adjust the control input by the adaptive law. Overall, there is improvement of the proposed adaptive stall flutter controller on the following aspects: (1) involving both of plunge acceleration and pitch rate in system output signal instead of only pitch rate for the output feedback control, (2) adding the cross term of $\dot{h}\dot{\theta}$ in the adaptive control law and (3) distributing the equal weigh on the control of plunged mode and pitch mode with $r_1 = -1$ and $r_2 = -1$. All those new control strategies can strengthen the coupling between pitching and plunging along the whole system trajectory and further improve the overall control performance. Hence, the proposed adaptive stall flutter controller is the effective and robust controller for the stall flutter suppression.

Moderate Stall Flutter Control versus Deep Stall Flutter Control

Figure illustrates the control responses of the adaptive stall flutter control of both moderate and deep stall flutter cases. For both stall flutter cases, the proposed adaptive stall flutter controller has the full controllability of both vibration modes, and it can suppress all LCOs with good dynamic performance in a short time within the saturation of the microtab.

However, the control effect in moderate stall flutter seems a little better than that of deep stall flutter. It can be found clearly that, compared with the moderate flutter case, the system responses of the deep stall flutter system tend to have a relatively larger overshoot, larger steady state error, and longer settling time, and the system needs more actuation of the microtab at around 1 second, when the system responses begin to converge. Therefore, it is more difficult to suppress the deep stall flutter than the moderate stall flutter. Note that the responses of the microtab deployment can quickly converge to zero after 1.5 seconds, which also confirms the fast actuation and the control capability of the microtab.

The control results can be verified by the results of microtab control analysis on stall flutter. As indicated in figure, the damping ratio of both plunge and pitch modes in stall flutter can be increased by the actuator of microtab with more significant improvement of plunge damping ratio than pitch damping ratio, which confirms the better control performance of plunge displacement with more smooth response in figure. At the same time, figure shows that damping ratios for moderate stall flutter can be improved in a better way than that for deep stall flutter, which verifies the conclusion of control results demonstrating that deep stall flutter control is tougher than the moderate stall flutter control.

The control results in figure are reasonable and can also be confirmed by physical explanation. Since the sharp change in aerodynamic loads for deep stall case leads to severe change of the coupling of vibration modes, the controller is more reluctant to the rapid change of the system dynamic and it will definitely need more power from the actuator to bond the strong coupling between vibration modes and gain the steady conditions.

Propeller and Airfoil Aerodynamics

A propeller is a device that transforms rotational power into linear thrust by creating a pressure between the two surfaces and acting upon a working fluid such as water or air. Airfoil aerodynamics refers to the cross-sectional shape of a wing that is used in various aeronautical applications. The diverse aspects of propeller and airfoil aerodynamics have been thoroughly discussed in this chapter.

Propeller

An aircraft propeller is an aerodynamic device which converts rotational energy into propulsive force creating thrust which is approximately perpendicular to its plane of rotation. The rotational energy can be produced by a piston or gas turbine engine or, in limited applications, by an electric motor. A propeller can be attached directly to the crankshaft of a piston engine, as is the case in many light aircraft, or it might be powered through a reduction gear box (RGB) attached to a piston or jet engine. In this case, the RGB converts the high rotation speed of the engine to one that is more appropriate for propeller operation. Propellers have two or more blades spaced evenly around the hub and are available in fixed pitch or in variable pitch configurations. More sophisticated propeller designs include those of the constant speed, contra-rotating and counter-rotating types.

Propeller Design

The cross section of a propeller is similar to that of a low drag wing and is subject to the same aerodynamic issues such as angle of attack, stall, drag and transonic air flow. There is a twist along the length of a propeller blade because the blade speed is much higher at the tip than it is at the root. The twist is necessary to maintain a more or less constant angle of attack along the length of the blade. Like a wing, behavior is degraded when it is not at its optimum angle of attack. To overcome this deficiency, many propellers use a variable pitch mechanism to adjust the blade pitch angle as the engine speed and aircraft velocity change.

Propeller design considerations include the number and shape of the blades but compromises are required. As an example, increasing the aspect ratio of the blade will reduce drag. However, as the amount of thrust that is produced by a propeller is proportional to the blade area, increasing the aspect ratio means that either longer blades or more blades are required to maintain equivalent thrust. Longer blades will approach transonic tip speed at a lower RPM than shorter ones and increasing the number of blades also results in an increase in the blade to blade interference effects.

The performance of a propeller diminishes greatly as the blade nears transonic speed. The relative airspeed at any point on a propeller is a vector sum of the tangential rotational speed of the propeller and the aircraft speed. As a result, the propeller blade tip will reach transonic speed well before the aircraft. At the critical speed, shock waves result in a significant increase in both drag and noise. Swept back, scimitar shaped propellers are used in some installations to increase the critical propeller speed and to reduce shock wave formation.

RCAF C130J six-bladed turboprop engines.

As the propeller moves through the air the static pressure is reduced ahead of each blade and at the same time at the blade face the flow is retarded resulting in an increase of static pressure. The changes in pressure around the rotating blades cause air to be drawn into the propeller disc and this results in a rearward movement of a column of air. The result being a forward thrust pulling the aircraft. All of this is in accordance with Newton's Third Law.

The amount of thrust generated by a propeller depends on the mass of the air and its acceleration toward the rear. Thus with equal RPM the smaller propeller delivers less thrust than the larger one.

Blade Twist and Pitch

The need for a changing blade angle from the hub to the tips stems from the fact that angular speed varies also and is greatest at the tips. Combine this with any forward speed the propeller may have, the relative airflow is also different from the hub to the tips. To keep thrust equal along the blade, they have a build in twist. The design is such that the blade is thick at the hub with a large blade angle and thin at the tip with a low blade angle.

During rotation and forward movement the propeller describes a rotational path, called a helix. If the propeller would move forward without giving thrust the distance of one revolution is called experimental mean pitch (angle between plane of rotation and the zero thrust angle of attack). The actual advance is the difference between experimental mean pitch and slip. Slip is the angle between the zero thrust and the actual angle of attack.

If the propeller rotates in a solid medium it would advance according to its pitch (angle between plane of rotation and the blade face, chord line,) also called geometric pitch.

Propeller Performance

If this was an ideal world the propeller would convert all power to thrust. But as this is not the case, losses occur in the slipstream and aerodynamic drag. Under normal conditions the propeller is able to convert about 85% of the brake horse power from the engine into thrust. Thus propeller efficiency is the ratio between thrust horsepower and brake horsepower.

Remember that:

Power = Force × Distance / Time (rate of doing work)

We can equate propeller efficiency as:

Thrust × TAS / Brake Horse Power

It follows that propeller efficiency is zero (0) under two conditions: when there is no forward speed (TAS) or when there is no thrust generated. With the aircraft at standstill (beginning of takeoff roll or taxi) the propeller has zero efficiency until it reaches its optimum forward speed for the propeller where maximum thrust is generated (maximum efficiency), increasing forward speed beyond that point will decrease efficiency (propeller with fixed blade angle). You could say that there is a relation between RPM and airspeed for fixed pitch propellers.

Variables Blades

It is obvious that a controllable propeller has a wider range of airspeeds where efficiency is at its maximum, until the governor reaches a position where the blades can no longer be adjusted, which are at the full ne and coarse pitch stops.

Blade Forces

During rotation the blades generate lift and drag. But with propeller we talk about thrust (lift) and propeller torque (drag). Another one: with a wing, drag must be overcome thrust to provide lift but with a propeller it is propeller torque that must be overcome by engine torque. Increasing power with the throttle increases engine torque, resulting in a higher RPM until propeller torque is equal to engine torque and RPM stabilizes.

If you would place an aircraft with a fixed pitch propeller into a shallow dive, as forward speed increases the relative airflow changes and the angle of attack is reduced. Resulting in a reduced thrust and propeller torque and as engine torque remained the same (there was no change in throttle setting) engine/propeller RPM will increase. The constant speed propeller would maintain the preset RPM.

Blade Effectiveness

Any propeller blade is the most effective between station 60% and 90% with a peak at 75%. It is this point (or station) where blade angle is usually reported in aircraft documentation.

Asymmetric Blade Effect or P-factor

This is most noticeable during high angles of attack of the airplane, takeoff rotation, and for tailwheel aircraft on their takeoff run. The main cause is that the propeller disc (normally at a right angle with the airflow) is tilted backwards and the relative airflow between the up going and down going blades is different.

The down going blades have a larger angle of attack (the distance traveled seems greater due to the forward movement of the aircraft) and produce more thrust, pulling the aircraft with a higher force than the up going blades. With an engine rotating clockwise (as seen from the cockpit) the aircraft wants to turn to the left as a result from this.

Twin Engine Aircraft

On twin engined propeller driven aircraft and with both propellers rotating in the same direction the arm (and distance) of the downward going blades towards the center of gravity are not the same. The engine where the distance is the smallest is called the critical engine. If that one fails the other engine will yaw the aircraft with more force (thrust x arm) than if the critical engine had failed.

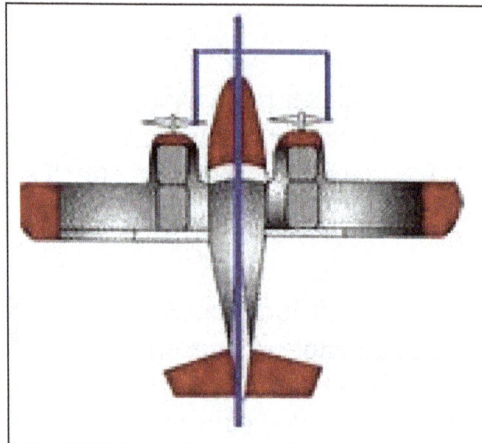

Blade Element Theory for Propellers

A relatively simple method of predicting the performance of a propeller (as well as fans or windmills) is the use of Blade Element Theory.

In this method the propeller is divided into a number of independent sections along the length. At each section a force balance is applied involving 2D section lift and drag with the thrust and torque produced by the section. At the same time a balance of axial and angular momentum is applied. This produces a set of non-linear equations that can be solved by iteration for each blade section. The resulting values of section thrust and torque can be summed to predict the overall performance of the propeller.

The theory does not include secondary effects such as 3-D flow velocities induced on the propeller by the shed tip vortex or radial components of flow induced by angular acceleration due to the rotation of the propeller. In comparison with real propeller results this theory will over-predict thrust and under-predict torque with a resulting increase in theoretical efficiency of 5% to 10% over measured performance. Some of the flow assumptions made also breakdown for extreme conditions when the flow on the blade becomes stalled or there is a significant proportion of the propeller blade in windmilling configuration while other parts are still thrust producing.

The theory has been found very useful for comparative studies such as optimising blade pitch setting for a given cruise speed or in determining the optimum blade solidity for a propeller. Given the above limitations it is still the best tool available for getting good first order predictions of thrust, torque and efficiency for propellers under a large range of operating conditions.

Blade Element Subdivision

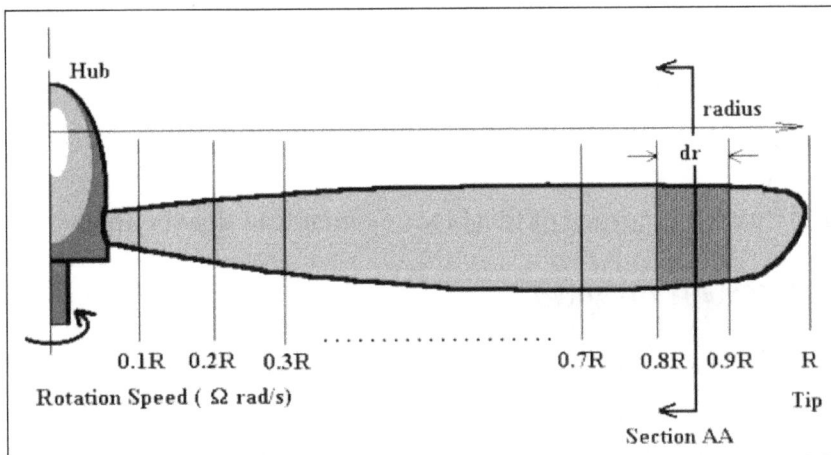

A propeller blade can be subdivided as shown into a discrete number of sections. For each section the flow can be analysed independently if the assumption is made that for each there are only axial and angular velocity components and that the induced flow input from other sections is negligible. Thus at section AA (radius = r) shown above, the flow on the blade would consist of the following components.

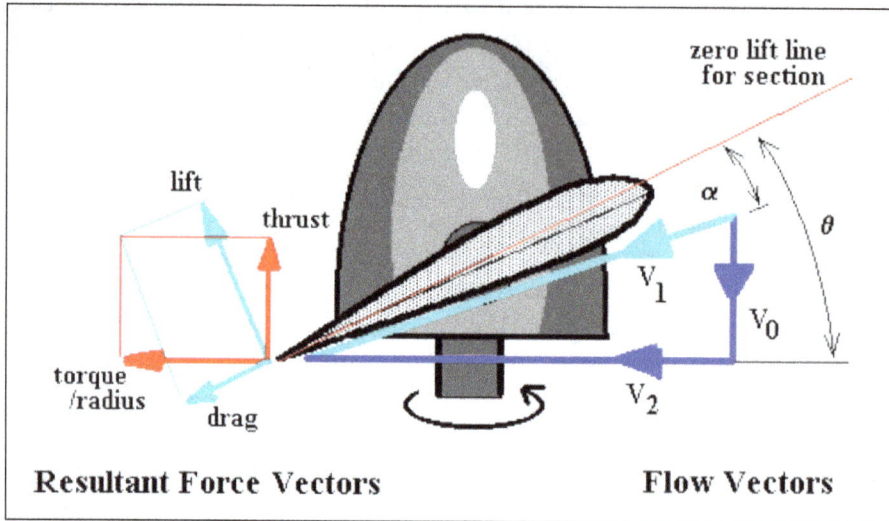

Resultant Force Vectors   **Flow Vectors**

- V_o axial flow at propeller disk, V_2 -- Angular flow velocity vector.

- V_1 section local flow velocity vector, summation of vectors V_o and V_2.

Since the propeller blade will be set at a given geometric pitch angle (θ) the local velocity vector will create a flow angle of attack on the section. Lift and drag of the section can be calculated using standard 2-D aerofoil properties. Propellers use a changed reference line: zero lift line not section chord line. The lift and drag components normal to and parallel to the propeller disk can be calculated so that the contribution to thrust and torque of the compete propeller from this single element can be found.

The difference in angle between thrust and lift directions is defined as:

$$\varphi = \theta - \alpha$$

The elemental thrust and torque of this blade element can thus be written as:

$$\Delta T = \Delta L \cos(\varphi) - \Delta D \sin(\varphi)$$

$$\frac{\Delta Q}{r} = \Delta D \cos(\varphi) + \Delta L \sin(\varphi)$$

Substituting section data (C_L and C_D for the given α) leads to the following equations,

$$\Delta L = C_L \frac{1}{2} \rho V_1^2 c . dr$$

$$\Delta D = C_D \frac{1}{2} \rho V_1^2 c . dr$$

per blade.

Where ρ is the air density, c is the blade chord so that the lift producing area of the blade element is c . dr. If the number of propeller blades is (B) then,

$$\Delta T = \frac{1}{2} \rho V_1^2 c \left(C_L \cos(\varphi) - C_D \sin(\varphi) \right) B . dr$$

$$\Delta Q = \frac{1}{2} \rho V_1^2 c \left(C_D \cos(\varphi) - C_L \sin(\varphi) \right) B . dr$$

Inflow Factors

A major complexity in applying this theory arises when trying to determine the magnitude of the two flow components V_0 and V_2. V_0 is roughly equal to the aircraft's forward velocity (V_∞) but is increased by the propeller's own induced axial flow into a slipstream. V_2 is roughly equal to the blade section's angular speed (Ωr) but is reduced slightly due to the swirling nature of the flow induced by the propeller. To calculate V_0 and V_2 accurately both axial and angular momentum balances must be applied to predict the induced flow effects on a given blade element. As shown in the following diagram, the induced components can be defined as factors increasing or decreasing the major flow components.

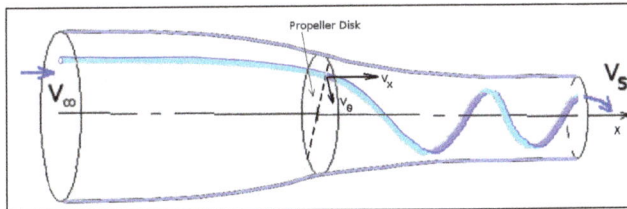

A typical streamtube of flow passing through section AA would have velocities:

$$V_\theta = b\Omega r \quad \text{and} \quad V_x = V_\infty (1+a)$$

So for the velocities V_0 and V_2,

$$V_0 = V_\infty + aV_\infty$$

where a is the axial inflow factor, and

$$V_2 = \Omega - bQr$$

b is the angular inflow factor (swirl factor).

The local flow velocity and the angle of attack for the blade section is thus:

$$V_1 = \sqrt{V_0^2 + V_2^2}$$

$$\alpha = \theta - \tan^{-1}\left(\frac{V_0}{V_2}\right)$$

Axial and Angular Flow Conservation of Momentum

The governing principle of conservation of flow momentum can be applied for both axial and circumferential directions. For the axial direction, the change in flow momentum along a stream-tube starting upstream, passing through the propeller at section AA and then moving off into the slipstream, must equal the thrust produced by this element of the blade.

To remove the unsteady effects due to the propeller's rotation, the stream-tube used is one covering the complete area of the propeller disk swept out by the blade element and all variables are assumed to be time averaged values.

ΔT = Change in Momemtum flow rate through tube at disk

= Mass flow rate in tube x Change in velocity along tube

$$= \rho.2\pi r.dr.V_0 \times (V_s - V_\infty)$$

By applying Bernoulli's equation and conservation of momentum, for the three separate components of the tube, from freestream to face of disk, from rear of disk to slipstream far downstream and balancing pressure and area versus thrust, it can be shown that the axial velocity at the disk will be the average of the freestream and slipstream velocities.

$$V_0 = \frac{V_\infty + V_s}{2}$$

Which means that:

$$V_s = V_\infty\left(1 + 2a\right)$$

Therefore:

$$\Delta T = 2\pi r \rho V_\infty (1+a)(V_\infty(1+2a) - V_\infty).dr$$
$$= 4\pi r \rho V_\infty^2 (1+a)a.dr$$

For angular momentum:

ΔQ = Change in angular momentum rate of flow in tube × radius

= Mass flow rate in tube x change in circumferential velocity × radius

$$= \rho.2\pi r.dr.V_0 \times (V_{\theta(slipstream)} - V_{\theta(freestream)}) \times r$$

By considering conservation of angular momentum in conjunction with the axial velocity change, it can be shown that the angular velocity in the slipstream will be twice the value at the propeller disk.

$$V_{\theta(slipstream)} = 2b\Omega r \quad \text{and} \quad V_{\theta(freestream)} = 0$$

Thus,

$$\Delta Q = 2\pi r \rho V_\infty (1+a)(2b\Omega r)r.dr$$
$$= 4\pi r^3 V_\infty (1+a)b\Omega.dr$$

Because these final forms of the momentum equation balance still contain the variables for element thrust and torque, they cannot be used directly to solve for inflow factors. However there now exists a nonlinear system of equations containing the four primary unknown variables ΔT, ΔQ, a, b, so an iterative solution to this system is possible.

Iterative Solution Procedure for Blade Element Theory

The method of solution for the blade element flow will be to start with some initial guess of inflow factors (a) and (b). Use these to find the flow angle on the blade, then use blade section properties to estimate the element thrust and torque. With these approximate values of thrust and torque previous equations can be used to give improved estimates of the inflow factors (a) and (b).

This process can be repeated until values for (a) and (b) have converged to within a specified tolerance.

It should be noted that convergence for this nonlinear system of equations is not guaranteed. It is usually a simple matter of applying some convergence enhancing techniques (ie Crank-Nicholson under-relaxation) to get a result when linear aerofoil section properties are used. When non-linear properties are used, ie including stall effects, then obtaining convergence will be significantly more difficult.

For the final values of inflow factor (a) and (b) an accurate prediction of element thrust and torque will be obtained from equations(all six equations mentioned above).

Propeller Thrust and Torque Coefficients and Efficiency

The overall propeller thrust and torque will be obtained by summing the results of all the radial blade element values.

$$T = \Sigma \Delta T_{(for\ all\ elements)} \quad and \quad Q = \Sigma \Delta Q_{(for\ all\ elements)}$$

The non-dimensional thrust and torque coefficients can then be calculated along with the advance ratio at which they have been calculated.

$$C_T = \frac{T}{\rho n^2 D^4}$$

and

$$C_Q = \frac{Q}{\rho n^2 D^5}$$

where n is the rotation speed of propeller in revs per second and D is the propeller diameter.

The efficiency of the propeller under these flight conditions will then be:

$$\eta_{prop} = \frac{J C_T}{2\pi C_Q}$$

where, advance ratio J is:

$$J = \frac{V_\infty}{nD}$$

Propeller Aerodynamic Process

An airplane moving through the air creates a drag force opposing its forward motion. If an airplane is to fly on a level path, there must be a force applied to it that is equal to the drag but acting forward. This force is called thrust. The work done by thrust is equal to the thrust times the distance it moves the airplane.

Work = Thrust × Distance

The power expended by thrust is equal to the thrust times the velocity at which it moves the airplane.

Power = Thrust × Velocity

If the power is measured in horsepower units, the power expended by the thrust is termed thrust horsepower.

The engine supplies brake horsepower through a rotating shaft, and the propeller converts it into thrust horsepower. In this conversion, some power is wasted. For maximum efficiency, the propeller must be designed to keep this waste as small as possible. Since the efficiency of any machine is the ratio of the useful power output to the power input, propeller efficiency is the ratio of thrust horsepower to brake horsepower. The usual symbol for propeller efficiency is the Greek letter η (eta). Propeller efficiency varies from 50 percent to 87 percent, depending on how much the propeller slips.

Pitch is not the same as blade angle, but because pitch is largely determined by blade angle, the two terms are often used interchangeably. An increase or decrease in one is usually associated with an increase or decrease in the other. Propeller slip is the difference between the geometric pitch of the propeller and its effective pitch. Geometric pitch is the distance a propeller should advance in one revolution with no slippage; effective pitch is the distance it actually advances. Thus, geometric or theoretical pitch is based on no slippage. Actual, or effective, pitch recognizes propeller slippage in the air. The relationship can be shown as:

Geometric pitch – Effective pitch = slip

Geometric pitch is usually expressed in pitch inches and calculated by using the following formula:

GP = 2 × π R × tangent of blade angle at 75 percent station

R = Radius at the 75 percent blade station

π = 3.14

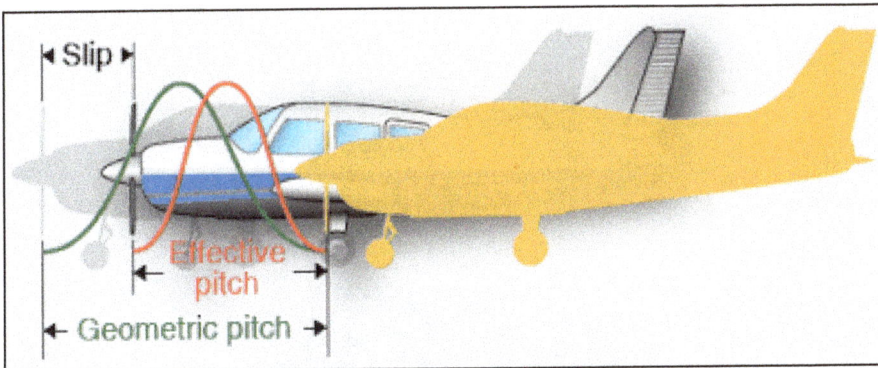

Effective pitch and geometric pitch.

Although blade angle and propeller pitch are closely related, blade angle is the angle between the face or chord of a blade section and the plane in which the propeller rotates. Blade angle, usually measured in degrees, is the angle between the chordline of the blade and the plane of rotation. The chordline of the propeller blade is determined in about the same manner as the chordline of an airfoil. In fact, a propeller blade can be considered as being composed of an infinite number of thin blade elements, each of which is a miniature airfoil section whose chord is the width of the propeller blade at that section. Because most propellers have a flat blade face, the chord line is often drawn along the face of the propeller blade.

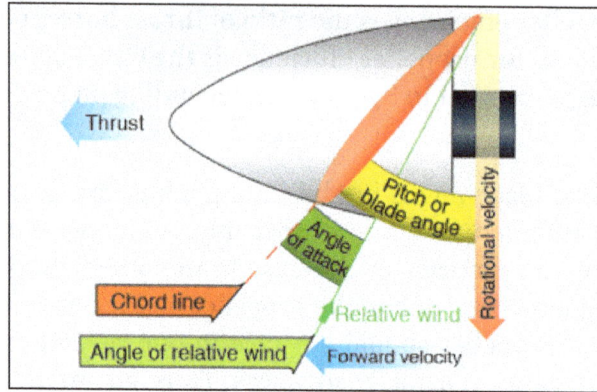

Propeller aerodynamic factors.

The typical propeller blade can be described as a twisted airfoil of irregular planform. Two views of a propeller blade are shown in figure. For purposes of analysis, a blade can be divided into segments that are located by station numbers in inches from the center of the blade hub. The cross-sections of each 6-inch blade segment are shown as airfoils in the right side of figure. Also identified in figure are the blade shank and the blade butt. The blade shank is the thick, rounded portion of the propeller blade near the hub and is designed to give strength to the blade. The blade butt, also called the blade base or root, is the end of the blade that fits in the propeller hub. The blade tip is that part of the propeller blade farthest from the hub, generally defined as the last 6 inches of the blade.

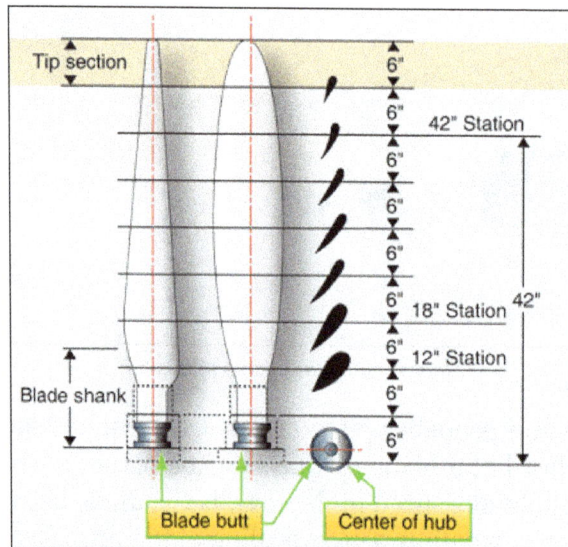

Typical propeller blade elements.

A cross-section of a typical propeller blade is shown in figure. This section or blade element is an airfoil comparable to a cross-section of an aircraft wing. The blade back is the cambered or curved side of the blade, similar to the upper surface of an aircraft wing. The blade face is the flat side of the propeller blade. The chord line is an imaginary line drawn through the blade from the leading edge to the trailing edge. The leading edge is the thick edge of the blade that meets the air as the propeller rotates.

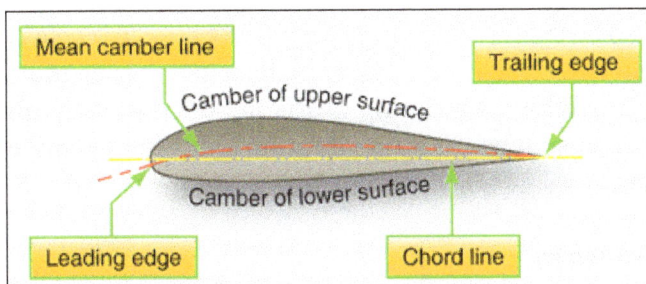

Cross-section of a propeller blade.

A rotating propeller is acted upon by centrifugal twisting, aerodynamic twisting, torque bending, and thrust bending forces. The principal forces acting on a rotating propeller are illustrated in figure.

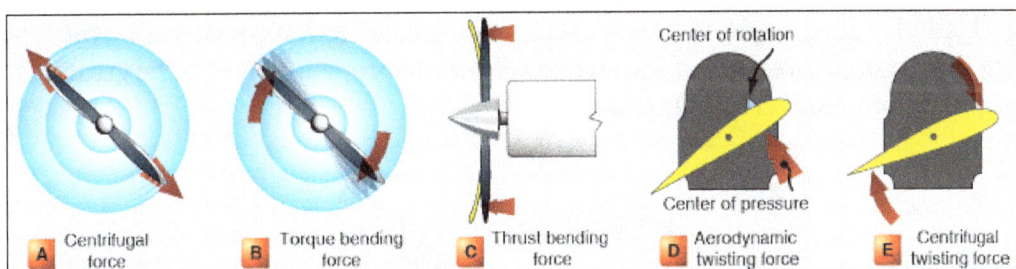

Forces acting on a rotating propeller.

Centrifugal force is a physical force that tends to throw the rotating propeller blades away from the hub. This is the most dominant force on the propeller. Torque bending force, in the form of air resistance, tends to bend the propeller blades in the direction opposite that of rotation. Thrust bending force is the thrust load that tends to bend propeller blades forward as the aircraft is pulled through the air. Aerodynamic twisting force tends to turn the blades to a high blade angle. Centrifugal twisting force, being greater than the aerodynamic twisting force, tends to force the blades toward a low blade angle.

At least two of these forces acting on the propellers blades are used to move the blades on a controllable pitch propeller. Centrifugal twisting force is sometimes used to move the blades to the low pitch position, while aerodynamic twisting force is used to move the blades into high pitch. These forces can be the primary or secondary forces that move the blades to the new pitch position.

A propeller must be capable of withstanding severe stresses, which are greater near the hub, caused by centrifugal force and thrust. The stresses increase in proportion to the rpm. The blade face is also subjected to tension from the centrifugal force and additional tension from the bending. For these reasons, nicks or scratches on the blade may cause very serious consequences. These could lead to cracks and failure of the blade.

A propeller must also be rigid enough to prevent fluttering, a type of vibration in which the ends of the blade twist back and forth at high frequency around an axis perpendicular to the engine crankshaft. Fluttering is accompanied by a distinctive noise, often mistaken for exhaust noise. The constant vibration tends to weaken the blade and eventually causes failure.

Aerodynamic Factors

To understand the action of a propeller, consider first its motion, which is both rotational and forward. Thus, as shown by the vectors of propeller forces in figure, a section of a propeller blade moves downward and forward. As far as the forces are concerned, the result is the same as if the blade were stationary and the air coming at it from a direction opposite its path. The angle at which this air (relative wind) strikes the propeller blade is called angle of attack (AOA). The air deflection produced by this angle causes the dynamic pressure at the engine side of the propeller blade to be greater than atmospheric pressure, creating thrust.

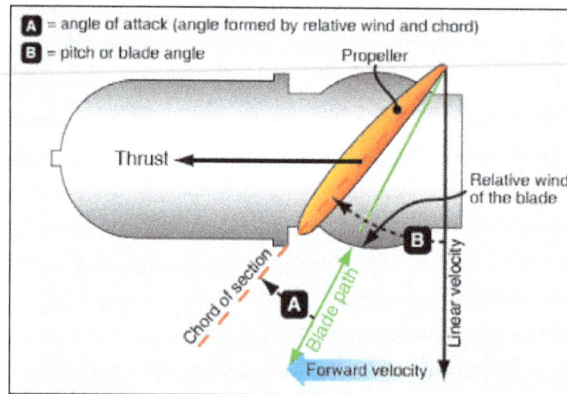

Propeller forces.

The shape of the blade also creates thrust because it is shaped like a wing. As the air flows past the propeller, the pressure on one side is less than that on the other. As in a wing, this difference in pressure produces a reaction force in the direction of the lesser pressure. The area above a wing has less pressure, and the force (lift) is upward. The area of decreased pressure is in front of a propeller which is mounted in a vertical instead of a horizontal position, and the force (thrust) is in a forward direction. Aerodynamically, thrust is the result of the propeller shape and the AOA of the blade.

Another way to consider thrust is in terms of the mass of air handled. In these terms, thrust is equal to the mass of air handled multiplied by the slipstream velocity minus the velocity of the airplane. Thus, the power expended in producing thrust depends on the mass of air moved per second. On the average, thrust constitutes approximately 80 percent of the torque (total horsepower absorbed by the propeller). The other 20 percent is lost in friction and slippage. For any speed of rotation, the horsepower absorbed by the propeller balances the horsepower delivered by the engine. For any single revolution of the propeller, the amount of air displaced (moved) depends on the blade angle, which determines the quantity or amount of mass of air the propeller moves. Thus, the blade angle is an excellent means of adjusting the load on the propeller to control the engine rpm. If the blade angle is increased, more load is placed on the engine, tending to slow it down unless more power is applied. As an airfoil is moved through the air, it produces two forces: lift and drag. Increasing propeller blade angle increases the AOA and produces more lift and drag; this action increases the horsepower required to turn the propeller at a given rpm. Since the engine is still producing the same horsepower, the propeller slows down. If the blade angle is decreased, the propeller speeds up. Thus, the engine rpm can be controlled by increasing or decreasing the blade angle.

The blade angle is also an excellent method of adjusting the AOA of the propeller. On constant-speed propellers, the blade angle must be adjusted to provide the most efficient AOA at all engine and airplane speeds. Lift versus drag curves, which are drawn for propellers as well as wings, indicate that the most efficient AOA is a small one varying from 2° to 4° positive. The actual blade angle necessary to maintain this small AOA varies with the forward speed of the airplane. This is due to a change in the relative wind direction, which varies with aircraft speed.

Fixed-pitch and ground-adjustable propellers are designed for best efficiency at one rotation and forward speed. In other words, they are designed to fit a given airplane and engine combination. A propeller may be used that provides the maximum propeller efficiency for takeoff, climb, cruising, or high speeds. Any change in these conditions results in lowering the efficiency of both the propeller and the engine. A constant-speed propeller, however, keeps the blade angle adjusted for maximum efficiency for most conditions encountered in flight. During takeoff, when maximum power and thrust are required, the constant-speed propeller is at a low propeller blade angle or pitch. The low blade angle keeps the AOA small and efficient with respect to the relative wind. At the same time, it allows the propeller to handle a smaller mass of air per revolution. This light load allows the engine to turn at high rpm and to convert the maximum amount of fuel into heat energy in a given time. The high rpm also creates maximum thrust. Although the mass of air handled per revolution is small, the engine rpm is high, the slipstream velocity (air coming off the propeller) is high, and, with the low airplane speed, the thrust is maximum.

After liftoff, as the speed of the airplane increases, the constant-speed propeller changes to a higher angle (or pitch). Again, the higher blade angle keeps the AOA small and efficient with respect to the relative wind. The higher blade angle increases the mass of air handled per revolution. This decreases the engine rpm, reducing fuel consumption and engine wear, and keeps thrust at a maximum.

For climb after takeoff, the power output of the engine is reduced to climb power by decreasing the manifold pressure and increasing the blade angle to lower the rpm. Thus, the torque (horsepower absorbed by the propeller) is reduced to match the reduced power of the engine. The AOA is again kept small by the increase in blade angle. The greater mass of air handled per second, in this case, is more than offset by the lower slipstream velocity and the increase in airspeed.

At cruising altitude, when the airplane is in level flight and less power is required than is used in takeoff or climb, engine power is again reduced by lowering the manifold pressure and increasing the blade angle to decrease the rpm. Again, this reduces torque to match the reduced engine power; for, although the mass of air handled per revolution is greater, it is more than offset by a decrease in slipstream velocity and an increase in airspeed. The AOA is still small because the blade angle has been increased with an increase in airspeed. Pitch distribution is the twist in the blade from the shank to the blade tip, due to the variation in speeds that each section of the blade is traveling. The tip of the blade is traveling much faster than the inner portion of the blade.

Propeller Controls and Instruments

Fixed pitch propellers have no controls and require no adjustments in flight. The constant-speed propeller has a propeller control in the center pedestal between the throttle and the mixture control. The two positions for the control are increase rpm (full forward) and decrease rpm (pulled aft). This control is directly connected to the propeller governor and, by moving the control, adjusts the tension on the governor speeder spring. This control can also be used to feather the propeller in some aircraft by moving the control to the full decrease rpm position. The two main instruments used with the constant-speed propeller are the engine tachometer and the manifold pressure gauge. Rotations per minute (rpm) are controlled by the propeller control and the manifold pressure is adjusted by the throttle.

Turboprop propeller controls.

Airfoil Aerodynamics

An aerofoil is the term used to describe the cross-sectional shape of an object that, when moved through a fluid such as air, creates an aerodynamic force. Aerofoils are employed on aircraft as wings to produce lift or as propeller blades to produce thrust. Both these forces are produce perpendicular to the air flow. Drag is a consequence of the production of lift/thrust and acts parallel to the airflow.

Other aerofoil surfaces includes tailplanes, fins, winglets, and helicopter rotor blades. Control surfaces (e.g. ailerons, elevators and rudders) are shaped to contribute to the overall aerofoil section of the wing or empennage.

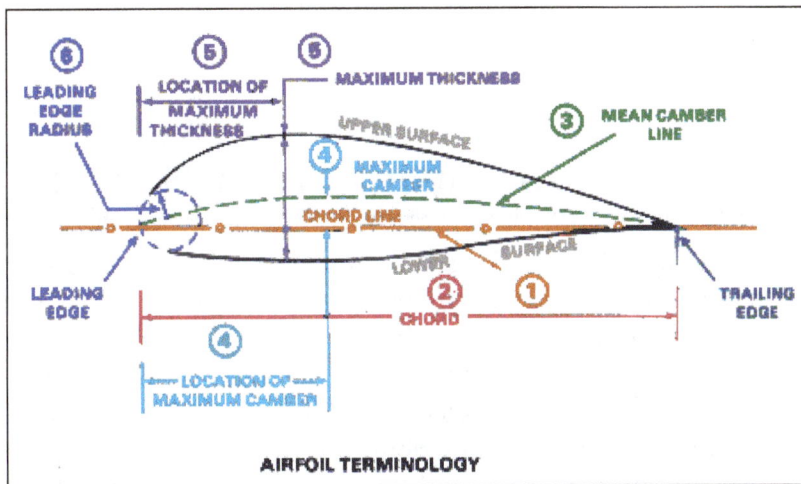

AIRFOIL TERMINOLOGY

Several terms are used to describe aerofoils:

- Leading Edge = Forward edge of the aerofoil.

- Trailing Edge = Aft edge of the aerofoil.

- Chord = Line connecting the leading and trailing edge. Denotes the length of the aerofoil.

- Mean Camber Line = Line drawn half way between the upper and lower surface of the aerofoil. Denotes the amount of curvature of the wing.

- Point of Maximum Thickness = Thickest part of the wing expressed as a percentage of the chord.

By altering each of the above features of an aerofoil, the designer is able to adjust the performance of the wing so that it is suitable for it's particular task. For example, a crop duster may have a thick, high camber wing that produces a large amount of lift at low speed. Alternatively, a jet would have a thin wing with minimal camber to allow it to cruise at high speeds.

Working of Aerofoil

The basic principle behind an aerofoil is described by bernoullis theorem. Basically this states that total pressure is equal to static pressure (due to the weight of air above) plus dynamic pressure (due to the motion of air).

Air that travels over the top surface of the aerofoil has to travel faster and thus gains dynamic pressure. The subsequent loss of static pressure creates a pressure difference between the upper and lower surfaces that is called lift and opposes the weight of an aircraft (or thrust that opposes drag).

As the angle of attack (the angle between the chord line and relative air flow) is increased, more lift is created. Once the critical angle of attack is reached (generally around 14 degrees) the aerofoil will stall.

Airfoil Aerodynamics using Panel Methods

Panel methods are numerical models based on simplifying assumptions about the physics and properties of the flow of air over an aircraft. The viscosity of air in the flow field is neglected, and the net effect of viscosity on a wing is summarized by requiring that the flow leaves the sharp trailing edge of the wing smoothly. The compressibility of air is neglected, and the curl of the velocity field is assumed to be zero (no vorticity in the flow field). Under these assumptions, the vector velocity describing the flow field can be represented as the gradient of a scalar velocity potential, , and the resulting flow is referred to as potential flow. A statement of conservation of mass in the flow field leads to Laplace's equation as the governing equation for the velocity potential. Laplace's equation is a widely studied linear partial differential equation. It also plays an important role in the theoretical development of several fields, including electrostatics and elastic membranes as well as fluid dynamics.

To solve the problem of potential flow over a solid object, Laplace's equation must be solved subject to the boundary condition that there is no flow across the surface of the object. This is usually referred to as the tangent-flow boundary condition. Additionally, the flow far from the object is required to be uniform. The results of solving Laplace's equation subject to tangent-flow boundary conditions provide an approximation of cruise conditions for an airplane.

Using a vector identity, the solution to this linear partial differential equation can be written in terms of an integral over the surface of the object. This boundary integral contains expressions for surface distributions of basic singular solutions to Laplace's equation. A linear combination of relatively simple singular solutions is also a solution to the differential equation. This superposition of simple solutions provides the complexity needed for satisfying boundary conditions for flow over objects of complex geometry. Commonly used singular solutions for panel methods are referred to as source, vortex, and doublet distributions. Analogies can be made to other fields of study. The velocity field induced by a point source is analogous to the electrostatic field induced by a point charge. A doublet would be positive and negative charges of equal strength in close proximity. The velocity induced by a line vortex is analogous to the magnetic field induced by a current-carrying wire.

The basic solution procedure for panel methods consists of discretizing the surface of the object with flat panels and selecting singularities to be distributed over the panels in a specified manner, but with unknown singularity-strength parameters. Since each singularity is a solution to Laplace's equation, a linear combination of the singular solutions is also a solution. The tangent-flow boundary condition is required to be satisfied at a discrete number of points called collocation points. This process leads to a system of linear algebraic equations to be solved for the unknown singularity-strength parameters. Details of the procedure vary depending on the singularities used and other details of problem formulation, but the end result is always a system of linear algebraic equations to be solved for the unknown singularity-strength parameters.

Panel methods are applicable to two- and three-dimensional flows. For flow over a two-dimensional object, the flat panels become straight lines, but can be thought of as infinitely long rectangular panels in the three-dimensional interpretation. For two-dimensional potential flow, the powerful technique of conformal mapping can also be used as a solution procedure. Conformal mapping provides exact solutions for certain airfoil shapes and is useful for validating numerical models.

Nomenclature and Basic Equations for Airfoil Aerodynamics

Much of the nomenclature associated with the theory of lift on an airfoil has made its way into everyday vocabulary, but some terms may be unfamiliar or have more specific meanings than occur in common usage.

The term airfoil is used to denote the cross section, or profile, of a three-dimensional wing. The chord line of an airfoil is the straight line from the leading edge of the airfoil to the sharp trailing edge; the length of this line is referred to as the chord of the airfoil and is denoted by c. The camber line of an airfoil is the locus of points midway between the upper and lower surfaces of the airfoil, measured perpendicular to the camber line. When describing an airfoil in dimensionless variables and in a local coordinate system, the chord of the airfoil is the segment of the x axis from 0 to 1. The angle of attack is the angle between the chord line of the airfoil and the uniform onset velocity, and is denoted by α.

In incompressible potential flow, the pressure is related to the fluid speed by Bernoulli's equation, $\frac{1}{2}\rho Q^2 + p = \frac{1}{2}\rho Q_\infty^2 + p_\infty$, where $\rho, Q,$ and p are the density, speed, and pressure at a point in the flow field, and the subscript ∞ refers to conditions far from the airfoil. The dimensionless measure of pressure is the pressure coefficient, defined by $C_p = \dfrac{p - p_\infty}{\dfrac{1}{2}\rho Q_\infty^2}$. Combining Bernoulli's equation and the definition for the pressure co-

efficient yields a simple equation for the pressure coefficient in terms of the local speed

of the fluid, $C_p = 1 - \dfrac{Q^2}{Q_\infty^2}$.

Using the aerodynamic sign convention, the circulation of the velocity field $\vec{Q}$ around a closed contour C is defined by the line integral $\Gamma = -\oint_C \vec{Q} \cdot d\vec{s}$. The lift force per unit length on an airfoil can be related to the circulation around the airfoil by the Kutta-Jou-kowski lift theorem $\ell = \rho Q_\infty \Gamma$. The aerodynamic sign convention used in the definition of circulation is chosen so that positive circulation leads to positive lift. The dimension-

less measure for lift on an airfoil is the two-dimensional lift coefficient, $c_l = \dfrac{\ell}{\dfrac{1}{2}\rho Q_\infty^2 c}$.

If a dimensionless circulation is defined by $\Gamma = \dfrac{r}{cQ_\infty}$, then the lift coefficient is simply

twice the dimensionless circulation. The Kutta condition summarizes the primary vis-cous effect of the flow on the airfoil and establishes the circulation around the airfoil by the simple statement that the flow leaves the sharp trailing edge of the airfoil smoothly.

Input Parameters

The example airfoil for illustrating the implementation of a panel method is a member of the NACA four-digit family of airfoils. Specify the identification number for the air-foil and the angle of attack in degrees.

$$\text{id} = 4412;\ \alpha = 10.0\ \text{Degree}$$

The first of the four digits in the identification number gives the maximum camber in percent chord, the second digit gives the location of maximum camber in tenths of chord, and the last two digits give the thickness in percent chord.

The discretization process is determined by a discretization number and one of three layout options (ConstantSpacing, CosineSpacing, or HalfCosineSpacing) providing two alternatives to constant spacing of discretization points. Specify small (to illustrate a step-by-step implementation of the example) and large (for computing results) discret-ization numbers, and a layout option.

$$\text{ns} = 3;\ \text{nl} = 100;\ \text{spacing} = \text{HalfCosineSpacing}$$

Finally, specify a small number used to ensure that collocation points are computed to be outside of the discretization panels representing the airfoil.

$$\varepsilon = 10^{-6}$$

Packages

Data Type for Handling Collections of Vectors

The package CartesianVectors defines a data type to simplify the manipulation of large collections of n-dimensional vectors while maintaining packed arrays for efficient computation using machine numbers. The data type CartesianVectors is represented in the format $\{vx \uparrow vy \uparrow vz\}$, where vx, vy, and vz are simple or nested lists of the components of the collection of vectors. The data type is designed to enable the manipulation of collections of vectors with notation commonly used for a single vector or for lists. Using a data type also simplifies pattern matching for valid input arguments for exported functions developed in other packages. The properties of the data type are defined by overloading existing Mathematica functions whenever possible. The package contains some exported functions including several functions specific to two-dimensional vectors.

As an example of using the data type, specify two collections of two-dimensional vectors, and compute the collection of displacement vectors from each vector in one group to every vector in the other group. This is a computation common to many n-body problems. The constructor for the data type is MakeCartesianVectors.

> rA = MakeCartesianVectors [{Array [xa, 3], Array [za, 3] }]

$$\{xa[1], xa[2], xa[3]\} \uparrow \{za[1], za[2], za[3]\}$$

> rB = MakeCartesianVectors [{Array [xb, 2], Array [zb, 2] }]

$$\{xb[1], xb[2]\} \uparrow \{zb[1], zb[2]\}$$

Compute the displacement vectors from each point in rB to all points in rA.

> (rAB = Outer [Plus, rA, − rB]) / / MatrixForm

$$\begin{pmatrix} xa[1] & -xb[1] & xa[1] & -xb[2] \\ xa[2] & -xb[1] & xa[2] & -xb[2] \\ xa[3] & -xb[1] & xa[3] & -xb[2] \end{pmatrix} \uparrow \begin{pmatrix} za[1] & -zb[1] & za[1] & -zb[2] \\ za[2] & -zb[1] & za[2] & -zb[2] \\ za[3] & -zb[1] & za[3] & -zb[2] \end{pmatrix}$$

Count the number of vectors in the collection of displacement vectors.

> Number of Vectors [rAB]

6

Airfoil Geometry

The package AirfoilGeometry provides functions to compute the geometry and discretization of airfoils in support of the construction of numerical models for potential flow over an airfoil. A list of x values used for discretization can be specified directly by the user or generated by the function NDiscretizeUnitSegment, which accepts a discretization number, n, as its input argument and divides the unit segment into n pieces. A layout option for this function allows constant spacing (default), cosine spacing, or half-cosine spacing. Cosine spacing provides finer discretization near the leading and trailing edges of the airfoil compared to constant spacing, and half-cosine spacing provides even finer discretization near the leading edge, but coarser discretization near the trailing edge compared to constant spacing. The function NACA4DigitAirfoil computes a list of thickness and camber properties at the x values, and the function AirfoilSurfacePoints computes the collection of vectors locating points on the surface of the airfoil from the list of thickness and camber properties. These points on the surface of the airfoil serve as panel end points for the discretized airfoil. Note that the result is expressed as the data type CartesianVectors as indicated by the arrow separating the lists of components.

rPanels =

AirfoilSurfacePoint [NACA4DigitAirfoil [id, NDiscretizeUnitSegment [ns, Layout $\rightarrow$ spacing]]]

$\{0.999833, 0.498824, 0.140789, 0., 0.127161, 0.501176, 1.00017\}$ $\uparrow$

$\quad \{-0.00124895, -0.0140383, - 0.289205, 0., 0.735357, 0.0918161, 0.00124895\}$

Compute a list of panel lengths. Note the use of Mathematica functions that have been overloaded for use with the data type CartesianVectors.

> lenghtPanels = PanelLengths [rPanels]

> $\{0.501173, 0358344, 0.143724, 0.146892, 0.374462, 0.507143\}$

Count the number of panels describing the discretized airfoil.

> nPanels = Length [lengthPanels]

> 6

The functions PanelPoints and PanelNormals are used to locate collocation points at mid-panel and outward facing unit normals to the panels. Note that collocation points are displaced a small distance, proportional to the panel length, in the direction of the outward unit normal to ensure that these points are outside the discretized airfoil. This is done in preparation for applying the tangent-flow boundary condition at collocation points.

unPanels = Pane1Normals [rPanels]

$\quad \{0.0255188, 0.415304, -0.201216, -0.50061, -0.0488176, 0.178583\}$ $\uparrow$

$\quad \{-0.999674, -0.999137, -0.979547, 0.865673, 0.998808, 0.983925\}$

rCollocation = PanelPoints [rPanels] + $\in$ MultiplyByList [lengthPanels, unPanels]

$$\{0.749328,\ 0.319806,\ 0.0703949,\ 0.0635802,\ 0.314168,\ 0.750571\}\uparrow$$

$$\{-0.00764412,\ -0.0214798,\ -0.0144604,\ 0.036768,\ 0.0826763,\ 0.046533\}$$

Figure shows the geometry of the discretized airfoil and the numbering convention for panels. The panels are straight-line segments joining points on the airfoil contour, and panel normals are shown at panel midpoints with the panel number near the head of the arrow. For an airfoil with thickness, the number of panels describing the airfoil is twice the discretization number. This numbering of panels is referred to as the clockwise convention. For a reference airfoil with no thickness (camber line), the number of panels is equal to the discretization number, and the convention is to number panels from leading edge to trailing edge. The airfoil shape is plotted in a local coordinate system with the origin at the leading edge of the airfoil and the x axis coincident with the chord line. Lengths are nondimensionalized using the chord of the airfoil, c.

gPanels = PlotAirfoil [rPanels];

gNormals = Graphics [Arrow [rCollocation, rCollocation + 0.15 unPanels]];

Show [gPanels, gNormals,

  Epilog $\rightarrow$ {Table [Text [i, rCollocation [[i]] + 0.17 unPanels [[i]]],

  {i, NumberofVectors [rcollocation] }] }, PlotRange $\rightarrow$ { {−0.25, 1.2}, {−0.25, 0.3} }]

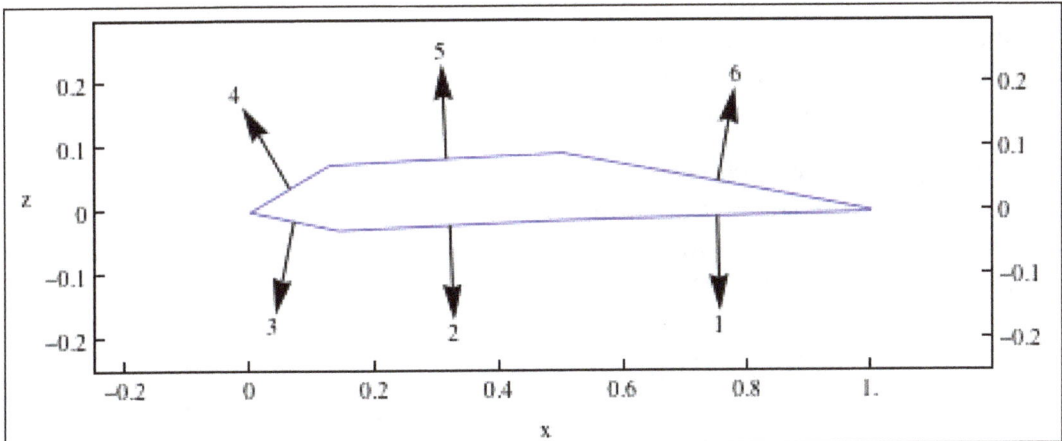

Panels and panel normals for NACA 4412 airfoil discretized to six panels.

These are defined by specifying points on the airfoil contour and rearranging the imported data for use with this software. The UIUC Airfoil Data Site, maintained by Michael Selig of the University of Illinois at Urbana-Champaign, contains specifications for over 1500 airfoils.

Two-dimensional Influence Coefficients

When computing the velocity field induced at a point due to a singularity located else-where, the velocity can be written as the product of a geometric term (called an influence coefficient) and a measure of the strength of the singularity. For example, consider the velocity induced at an arbitrary field point r_f due to a point source located at the origin of the coordinate system.

rf $= \{x, z\}$

$\{x, z\}$

rs $= \{0, 0\}$

$\{0, 0\}$

Compute the velocity, velocity-potential, and stream-function influence coefficients at the field point due to the singularity using the package functions ICSourcePoint, ICPhiSourcePoint, and ICPsiSourcePoint.

{ic, icϕ, icψ} = {ICSourcePoint [rf, rs], ICPhiSourcePoint [rf, rs], ICPsiSourcePoint [rf, rs]}

$$\left\{ \left\{ \frac{x}{2\pi\left(x^2+z^2\right)}, \frac{z}{2\pi\left(x^2+z^2\right)} \right\}, \frac{\text{Log}\left[x^2-z^2\right]}{4\pi}, \frac{\text{ArcTan}\left[x,z\right]}{2\pi} \right\}$$

The velocity, velocity-potential, or stream-function at r_f would be obtained by multiplying the appropriate influence coefficient by the strength of the source. Influence coefficients can also be thought of as the velocity, velocity-potential, or stream-function induced by a singularity of unit strength.

The package *InfluenceCoefficients* contains over thirty functions for velocity, velocity-potential, and stream-function influence coefficients for source, vortex, and doublet singularities commonly used in two-dimensional panel methods. They serve as a tool box for constructing numerical models for two-dimensional potential flow.

Potential-flow Model using Vortex Panels of Linearly Varying Strength

Step-by-Step Model Formulation using Coarse Discretization

The singularity element chosen for this model is the vortex panel of linearly varying strength, which provides a circulation density along the ith panel of the form, $\gamma_i\left(x_i\right) = \gamma_{0i} + s_i x_i$ in local coordinates, where x_i is the distance from the "beginning" of the panel. Each singularity panel involves two unknown constants, γ_{0i} and s_i.

The boundary condition that the velocity be everywhere tangent to the airfoil contour is discretized to require that the velocity component normal to each panel at the collocation point be zero. Since each vortex panel introduces two unknown strength parameters, application of the tangent-flow boundary condition provides n_{panels} equations and $2n_{panels}$ unknowns, where n_{panels} is the number of panels describing the geometry of the discretized airfoil. Continuity of circulation density from one panel to the next and the Kutta condition provide n_{panels} additional equations to complete a system of $2n_{panels}$ linear algebraic equations and $2n_{panels}$ unknowns. The system of equations can be put into standard form. The terms involving unknowns are collected on the left-hand side of the system of equations and the known quantities are collected on the right-hand side. The result can be written in block-matrix form as:

$$\begin{pmatrix} a_{11} & a_{12} \\ a_{21} & a_{22} \end{pmatrix} \begin{pmatrix} \gamma_0 \\ s \end{pmatrix} \begin{pmatrix} -Q_n \\ 0 \end{pmatrix}.$$

The symbols γ_0 and s represent lists of the unknown constant and linear strength parameters for the vortex panels: a_{11} represents the projection of the panel influence coefficients associated with γ_0 on the unit normal vectors, a_{12} represents the projection of the panel influence coefficients associated with s on the unit normal vectors, a_{21} Math and a_{22} represent terms in the equations imposing continuity of circulation density between panels and the Kutta condition, and Q_n is the projection of the free-stream velocity on unit normals at collocation points.

Use the block-matrix form to write the system of equations as $a_{11}\gamma_0 + a_{12}s = -Q_n$ and $a_{21}\gamma_0 + a_{22}s = 0$. Solve the latter system for the list of slope strengths, $s = -a_{22}^{-1}a_{21}\gamma_0$. Substitute this into the former system of equations to eliminate the slope-strength parameters. The resulting system of equations can be written as $\left(a_{11} - a_{12}\,a_{22}^{-1}a_{21}\right)\gamma_0 = -Q_n$. This system of equations can be solved for the list of strength parameters γ_0, and then the transformation is used to compute the list of slope parameters s. All variables in the following formulation and solution are dimensionless.

Compute the matrix of velocity influence coefficients and project them on the panel normals.

    (ico, ics) = ICVortexLinear [rCollection, rPanels]; a11 = ico . unPanels;

    a12 = ics . unPanels;

Write the equations expressing the continuity of circulation density between panels and the Kutta condition. The equation expressing the Kutta condition is written to accommodate the different numbering conventions for airfoils with thickness and reference airfoils without thickness.

a21 = If [nPanels = 2 ns, Module [{d}, d = DiagonalMatrix [Table [1.0, {nPan-els}]];

ReplacePart [−d + RotateLeft [d, {0, 1}], 1., {1, 1}]],

Module {d}, d = DiagonalMatrix [Table [1.0, {nPanels}]] ;

ReplacePart [−d + RotateLeft [d, {0, 1}, 0, {1, 1}]]];

a22 = RotateRight [DiagonalMatrix [lengthPanels]];

Form the coefficient matrix for the system of linear algebraic equations to be solved for unknown strength parameters.

a = ArrayFlatten $\left[\{ \{ a11, a12\}, \{a21, a22 \} \} \right]$;

Display the matrix in reduced precision to illustrate the coefficient matrix for the full system of equations.

Chop [NumberFrom[MatrixForm [a], {3, 2}, NumberPadding → {"o", "o"}, Number-Signs → {"−", "+"}, SignPadding → True]]

$$
\begin{pmatrix}
+0.00 & +0.14 & +0.03 & +0.03 & +0.14 & -0.08 & +0.08 & +0.02 & +0.00 & +0.00 & +0.03 & -0.08 \\
-0.21 & +0.00 & +0.09 & +0.09 & +0.00 & -0.20 & -0.06 & +0.06 & +0.01 & +0.01 & -0.02 & -0.04 \\
-0.12 & -0.28 & +0.00 & +0.16 & -0.19 & -0.12 & -0.03 & -0.07 & +0.02 & +0.00 & -0.03 & -0.03 \\
+0.10 & +0.16 & -0.16 & +0.00 & +0.28 & +0.11 & +0.03 & +0.03 & -0.02 & +0.02 & +0.04 & +0.02 \\
+0.19 & -0.00 & -0.08 & -0.09 & +0.00 & +0.21 & +0.06 & -0.02 & -0.01 & -0.01 & +0.06 & +0.04 \\
+0.08 & -0.13 & -0.03 & -0.03 & -0.14 & +0.00 & -0.03 & -0.02 & -0.00 & -0.00 & -0.03 & +0.08 \\
+1.00 & +0.00 & +0.00 & +0.00 & +0.00 & +1.00 & +0.00 & +0.00 & +0.00 & +0.00 & +0.00 & +0.51 \\
+1.00 & -1.00 & +0.00 & +0.00 & +0.00 & +0.00 & +0.50 & +0.00 & +0.00 & +0.00 & +0.00 & +0.00 \\
+0.00 & +1.00 & -1.00 & +0.00 & +0.00 & +0.00 & +0.00 & +0.36 & +0.00 & +0.00 & +0.00 & +0.00 \\
+0.00 & +0.00 & +1.00 & -1.00 & +0.00 & +0.00 & +0.00 & +0.00 & +0.14 & +0.00 & +0.00 & +0.00 \\
+0.00 & +0.00 & +0.00 & +1.00 & -1.00 & +0.00 & +0.00 & +0.00 & +0.00 & 0.15 & +0.00 & +0.00 \\
+0.00 & +0.00 & +0.00 & +0.00 & +1.00 & -1.00 & +0.00 & +0.00 & +0.00 & 0.00 & +0.37 & +0.00
\end{pmatrix}
$$

The upper half of the matrix represents normal-component influence coefficients. The first row of the lower half of the matrix represents terms in an equation implementing the Kutta condition and sums the circulation density at the beginning of the first panel and the circulation density at the end of the last panel. Setting this sum to zero imposes zero circulation at the trailing edge of the airfoil. The remaining rows in the lower half of the matrix are coefficients of terms in the equations requiring that the circulation density at the end of one panel be equal to the circulation density at the beginning of the next panel, $\gamma_{0j} + \delta_j s_j = \gamma_{0j+1}$, where δ_j denotes the length of the jth panel.

Define the transformation matrix to compute the list of slope parameters (s) from the list of constant parameters (γ_0).

$\qquad$ sFrom(γ_0) = – Inverse [a22] . a21;

Compute the free-stream velocity at collocation points.

$\qquad$ qInf = UniformFlow [nPanels, α];

Compute the components of the uniform flow normal to panels at collocation points.

$\qquad$ qnInf = qInf . unpanels;

Solve the system of equations for the list of constant parameters.

$\qquad$ (γ_0) = LinearSolve [all + a12 . sFrom(γ_0), –qnInf]

$\qquad$ $\{-1.26787,\ -0.814616,\ -0.685836,\ 1.19696,\ 1.76145,\ 1.41825\}$

Use the transformation matrix to compute the list of slope parameters.

$\qquad$ s = sFrom(γ_0)

$\qquad$ $\{0.90439,\ 0.359375,\ 13.0997,\ 3.8427,\ -0.91643,\ -0.296589\}$

The lift coefficient for the airfoil can be computed using the Kutta-Joukowski theorem. Recall that the distribution of circulation on a panel in local panel coordinates can be written as

$$\gamma_i(x_i) = \gamma_{0i} + s_i x_i,$$

where, x_i denotes the distance from the leading edge of the panel. The contribution of each panel to the lift is computed and the results summed over all panels.

In terms of the dimensionless variables used in this example, the contribution by each panel to the lift coefficient is just twice the net circulation associated with the panel, which is obtained by integrating the linear circulation density function, $\Delta c_{li} = 2 \int_0^{b_1} \gamma_i(x_i)\,dx_i = 2\gamma_{0i}\delta_i + s_i\delta_i^2$. Compute contributions of each panel to the airfoil lift coefficient.

$\qquad$ $\{\Delta\,\text{clC},\ \Delta\,\text{clL}\}$ = $\{2.0\ \gamma_0\,\text{lenghPanels},\ s\ \text{lengthPanels}^2\}$

$\qquad$ $\{\{-1.27085,\ -0.58326,\ -0.197148,\ 0.351648,\ 1.31919,\ 1.43855\},$

$\qquad$ $\{0.227159,\ 0.0461477,\ 0.270611,\ 0.0829194,\ -0.128503,\ -0.0762807\}\}$

Sum the two terms for each panel to obtain the list of contributions of each panel to the lift coefficient.

$\Delta cl = \Delta clC + \Delta clL$

$\{-1.04369, -0.537678, 0.0734631, 0.434568, 1.19069, 1.36226\}$

Sum the panel contributions to obtain the airfoil lift coefficient.

cl = Total $[\Delta cl]$

1. 47962

The computations illustrate the process of model implementation using a coarse discretization so that intermediate results can be viewed; however, the discretization is too coarse to provide useful results.

Remove names from computer memory, except those with values needed which presents an example computation of the pressure distribution and lift coefficient for a specified airfoil using a larger discretization number.

Apply [Remove, Complement [Name ["Global $*$"], {"id", "nl", "spacing", "α", "$\in$"}]]

Numerical Model for Fine Discretization

$\{\gamma 0, s\}$ = Module [{a, a11, a12, a21, a22, ts γ, qInf, qnInf, γ, δ},

rp = AirfoilSurfacePoint [

NACA4DigitAirfoil [id, NDiscretizeUnitSegment [n1, Layout $\rightarrow$ spacing]]];

δ = Drop [rP – RotateRight [rP],]; 1p = $\sqrt{\delta.\delta}$; np = Length [1p]; un = PanelNormals[rP];

rC = PanelPoints [rP] + $\in$ MultiplyByList [1p, un]; {ic0, ics} = ICVortexLinear [rC, rP];

a11 = ic0 . un; a12 = ics.un;

a21 = If [np = 2 nl, Module [{d}, d = DiagonalMatrix [Table [1.0, {np}]];

ReplacePart [–d + RotateLeft [d, {0, 1},1., {1,1}]]],

Module [{d}, d = DiagonalMatrix [Table [1.0, {np}]];

ReplacePart [–d + RotateLeft [d, {0, 1}, 0, {1, 1}]]]]; a22 = RotateRight [DiagonalMatrix [1p]];

ts γ = –Inverse [a22] . a21; a = a11 + a12.ts γ; qInf = UniformFlow [np, α]; qnInf = qInf . un;

γ = LinearSolve [a, – qnInf]; {γ, ts γ . γ} ;];

The results of this computation are the singularity strength parameters for all panels.

This model implementation has been validated by computing the results for a van de Vooren airfoil for which an exact solution is known by the method of conformal mapping. Also, convergence and timing studies have been performed and are available as online help documents in the software collection.

Pressure and Lift Coefficients

Use previously computed influence coefficients to determine the pressure coefficient at collocation points using Bernoulli's equation.

cp = Module [{q, qInf}, qInf = UniformFlow [np, α]; q = qInf + ino . γ 0 + ics . s; 1 − q . q]:

Lift and pitching moments can be computed from the pressure distribution. For example, the lift coefficient is computed by approximating the integral, $c_1 = -\oint C_p \hat{n} \cdot \hat{\ell} \, ds$, where the integral is over the airfoil contour, C_p is the pressure coefficient, $\hat{n}$ is the outward unit normal to the airfoil surface, and $\hat{\ell}$ is a unit vector perpendicular to the free-stream velocity in the direction of positive lift. The integral is approximated by considering the pressure coefficient constant over each panel, computing the contribution to lift of each panel, and summing the results.

clFromCp = Module[{ul},

ul = MakeCartesianVectors [{−Table [Sin[α], {np}], Table [Cos[α], {np}] }];

Δ clP = − cp (ul . un) lp;

Total [Δ clP]]

1. 70321

The lift can also be computed from the circulation distribution.

cl = Module [{ Δ cl, Δ clc, Δ clL}, Δ clc = 2.0 γ 0 1p; Δ clL = s l p²; Δ cl = Δ clclc + Δ clL; Total [Δ cl]]

1. 71006

Figure shows the surface pressure distribution on the airfoil in the conventional manner for such plots. Useful information from such plots include the locations of the stagnation point and the point of minimum pressure, and the severity of the positive pressure gradient on the upper surface.

xC = Components [rC] [] ;

PlotSurfacePressureCoefficient [xC, cp,

Epilog $\rightarrow$ Inset [PlotAirfoil [rP, Frame $\rightarrow$ False, PlotRange $\rightarrow$ {{0, 1}, Automatic}], {0,0}, {0,0}, 1]]

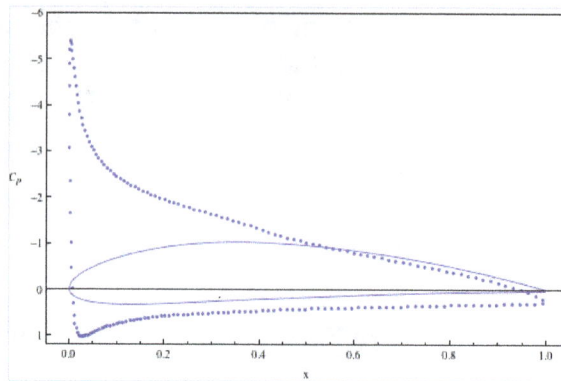

Pressure coefficient for a NACA 4412 airfoil, $\alpha = 10.0^\infty$ discretization 200 panels with HalfCosineSpacing.

Propellers used on General Aviation Aircraft

An increasing number of light aircraft are designed for operation with governor-regulated, constant-speed propellers. Significant segments of general aviation aircraft are still operated with fixed-pitch propellers. Light sport aircraft (LSA) use multiblade fixed-pitch composite propellers on up to medium size turbo prop aircraft with reversing propeller systems. Larger transport and cargo turbo prop aircraft use propeller systems with dual or double-acting governors and differential oil pressure to change pitch.

Fixed-pitch Wooden Propellers

Although many of the wood propellers were used on older airplanes, some are still in use. The construction of a fixed pitch, wooden propeller is such that its blade pitch cannot be changed after manufacture. The choice of the blade angle is decided by the normal use of the propeller on an aircraft during level flight when the engine performs at maximum efficiency. The impossibility of changing the blade pitch on the fixed-pitch propeller restricts its use to small aircraft with low horsepower engines in which maximum engine efficiency during all flight conditions is of lesser importance than in larger aircraft. The wooden, fixed-pitch propeller is well suited for such small aircraft because of its light weight, rigidity, economy of production, simplicity of construction, and ease of replacement.

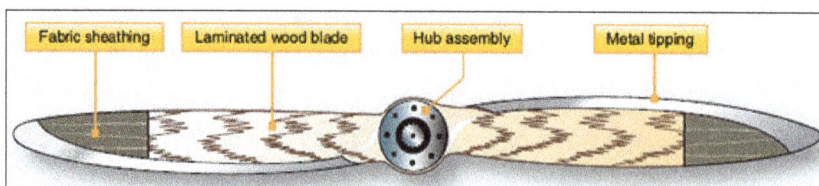

Fix-pitch wooden propeller assembly.

A wooden propeller is not constructed from a solid block, but is built up of a number of separate layers of carefully selected and well-seasoned hardwoods. Many woods, such as mahogany, cherry, black walnut, and oak, is used to some extent, but birch is the most widely used. Five to nine separate layers are used, each about 3/4 inch thick. The several layers are glued together with a waterproof, resinous glue and allowed to set. The blank is then roughed to the approximate shape and size of the finished product. The roughed-out propeller is then allowed to dry for approximately one week to permit the moisture content of the layers to become equalized. This additional period of seasoning prevents warping and cracking that might occur if the blank were immediately carved. Following this period, the propeller is carefully constructed. Templates and bench protractors are used to assure the proper contour and blade angle at all stations.

After the propeller blades are finished, a fabric covering is cemented to the outer 12 or 15 inches of each finished blade. A metal tipping is fastened to most of the leading edge and tip of each blade to protect the propeller from damage caused by flying particles in the air during landing, taxiing, or takeoff. Metal tipping may be of terneplate, Monel metal, or brass. Stainless steel has been used to some extent. It is secured to the leading edge of the blade by countersunk wood screws and rivets. The heads of the screws are soldered to the tipping to prevent loosening, and the solder is filed to make a smooth surface. Since moisture condenses on the tipping between the metal and the wood, the tipping is provided with small holes near the blade tip to allow this moisture to drain away or be thrown out by centrifugal force. It is important that these drain holes be kept open at all times. Since wood is subject to swelling, shrinking, and warping because of changes of moisture content, a protective coating is applied to the finished propeller to prevent a rapid change of moisture content. The finish most commonly used is a number of coats of water-repellent, clear varnish. After these processes are completed, the propeller is mounted on a spindle and very carefully balanced.

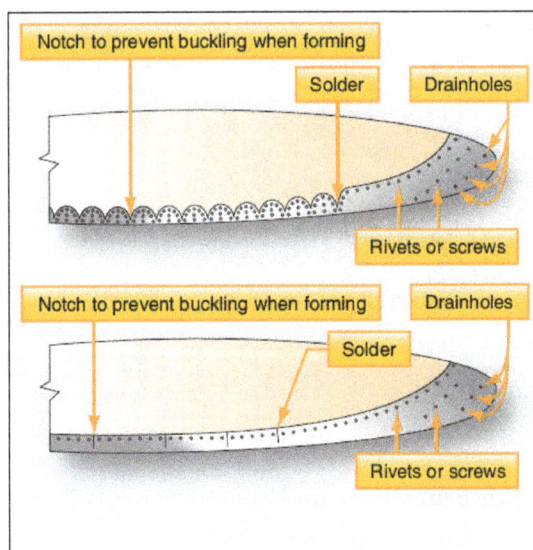

Installation of metal sheath and tipping.

Several types of hubs are used to mount wooden propellers on the engine crankshaft. The propeller may have a forged steel hub that fits a splined crankshaft; it may be connected to a tapered crankshaft by a tapered, forged steel hub; or it may be bolted to a steel flange forged on the crankshaft. In any case, several attaching parts are required to mount the propeller on the shaft properly.

Hubs fitting a tapered shaft are usually held in place by a retaining nut that screws onto the end of the shaft. On one model, a locknut is used to safety the retaining nut and to provide a puller for removing the propeller from the shaft. This nut screws into the hub and against the retaining nut. The locknut and the retaining nut are safetied together with lock-wire or a cotter pin.

Front and rear cones may be used to seat the propeller properly on a splined shaft. The rear cone is a one-piece bronze cone that fits around the shaft and against the thrust nut (or spacer) and seats in the rear-cone seat of the hub.

The front cone is a two-piece, split-type steel cone that has a groove around its inner circumference so that it can be fitted over a flange of the propeller retaining nut. Then, the retaining nut is threaded into place and the front cone seats in the front cone hub. A snap ring is fitted into a groove in the hub in front of the front cone so that when the retaining nut is unscrewed from the propeller shaft, the front cone acts against the snap ring and pulls the propeller from the shaft.

One type of hub incorporates a bronze bushing instead of a front cone. When this type of hub is used, it may be necessary to use a puller to start the propeller from the shaft. A rearcone spacer is sometimes provided with the splined-shaft propeller assembly to prevent the propeller from interfering with the engine cowling. The wide flange on the rear face of some types of hubs eliminates the use of a rear-cone spacer.

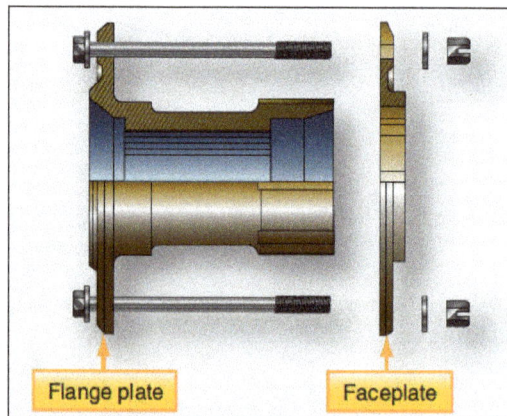

Hub assembly.

One type of hub assembly for the fixed-pitch, wooden propeller is a steel fitting inserted in the propeller to mount it on the propeller shaft. It has two main parts: the faceplate and the flange plate. The faceplate is a steel disk that forms the forward face of the hub.

The flange plate is a steel flange with an internal bore splined to receive the propeller shaft. The end of the flange plate opposite the flange disk is externally splined to receive the faceplate; the faceplate bore has splines to match these external splines. Both faceplate and flange plates have a corresponding series of holes drilled on the disk surface concentric with the hub center. The bore of the flange plate has a 15° cone seat on the rear end and a 30° cone seat on the forward end to center the hub accurately on the propeller shaft.

Metal Fixed-pitch

Propellers Metal fixed-pitch propellers are similar in general appearance to a wooden propeller, except that the sections are usually thinner. The metal fixed-pitch propeller is widely used on many models of light aircraft and LSA. Many of the earliest metal propellers were manufactured in one piece of forged Duralumin. Compared to wooden propellers, they were lighter in weight because of elimination of blade-clamping devices, offered a lower maintenance cost because they were made in one piece, provided more efficient cooling because of the effective pitch nearer the hub, and, because there was no joint between the blades and the hub, the propeller pitch could be changed, within limits, by twisting the blade slightly by a propeller repair station.

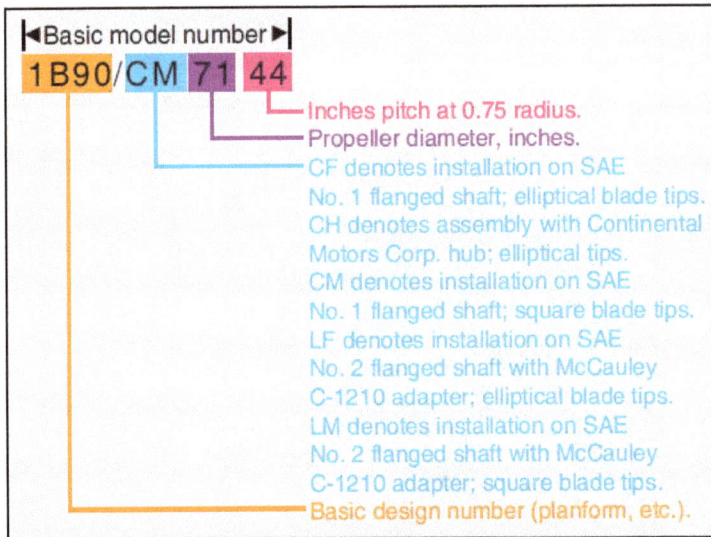

Complete propeller model numbers.

Propellers of this type are now manufactured as one-piece anodized aluminum alloy. They are identified by stamping the propeller hub with the serial number, model number, Federal Aviation Administration (FAA) type certificate number, production certificate number, and the number of times the propeller has been reconditioned. The complete model number of the propeller is a combination of the basic model number and suffix numbers to indicate the propeller diameter and pitch. An explanation of a complete model number, using the McCauley 1B90/CM propeller.

References

- Propeller: skybrary.aero, Retrieved 18 July, 2019

- Aircraft-propeller-2: experimentalaircraft.info, Retrieved 20 January, 2019

- Blade-element-propeller-theory, propulsion: aerodynamics4students.com, Retrieved 14 May, 2019

- Propeller-aerodynamic-process-airplane: aircraftsystemstech.com, Retrieved 19 April, 2019

- Aviation-aerofoil: aviationknowledge.wikidot.com, Retrieved 16 June, 2019

- Airfoil-aerodynamics-using-panel-methods: mathematica-journal.com, Retrieved 19 June, 2019

- Propellers-used-on-general-aviation-aircraft: flight-mechanic.com, Retrieved 16 March, 2019

Measurement Tools for Aerodynamic Pressure

The flight instruments depend on direct measurement of aerodynamic pressure to predict the altitude, airspeed and climb rate of the aircraft. It includes barometric altimeter, radio altimeter, vertical speed indicator, pilot-static probe, etc. This chapter closely examines these measurement tools of aerodynamic pressure to provide an extensive understanding of the subject.

Altimeter

An altimeter is a device that measures altitude, the distance of a point above sea level. Altimeters are important navigation instruments for aircraft and spacecraft pilots who monitor their height above the Earth's surface. Skydivers and mountaineers also use altimeters to pinpoint their location in the sky or on the ground.

The most common types of altimeters are barometric. They determine altitude by measuring air pressure. As altitude increases, air pressure decreases. This is because the density of air is lower (thinner) at high altitudes. It exerts less pressure on the Earth below.

An altimeter's readings change as elevation changes. The atmospheric pressure on Denali, Alaska, is about half that of Honolulu, Hawai'i. Denali, also known as Mount McKinley, is the highest peak in North America. Honolulu is a city at sea level.

Altitude readings can also change due to weather, as air pressure decreases during storms.

A simple barometric altimeter includes a sealed metal chamber, a spring, and a pointer that shows altitude in meters or feet. The chamber expands as air pressure decreases and contracts as it increases, bending the spring and moving the pointer. An altimeter can be mounted on an aircraft's instrument panel or worn on a person's wrist.

Other Types of Altimeters

Not all altimeters depend on air pressure. The Global Positioning System (GPS), for instance, can provide altitude as part of an area's location by triangulating signals from different satellites.

Radar and laser altimeters, found on some aircraft and spacecraft, work similarly to sonar measurements of the seafloor. These altimeters send a radio or laser signal toward the surface and measure the time it takes for the signal to bounce back. The time it takes for the signal to bounce back (or echo) to the aircraft is then translated to an elevation.

When used in satellites, radar and laser altimeters are able to combine altitude measurements to create accurate topographic maps of both land and ocean surfaces. The radar altimeter aboard the TOPEX/Poseidon satellite, for example, measured the surface topography of 95 percent of the ice-free ocean. Developed by NASA and CNES, the French space agency, TOPEX/Poseidon's radar altimeter was accurate to within 2 centimeters (less than 1 inch). Coupled with another satellite, Jason-1, TOPEX/Poseidon graphed the rise in global sea levels, providing evidence of the connection between global climate change and sea level rise.

An altimeter measures altitude.

Working of an Altimeter

Conventional aircraft altimeters work by measuring the atmospheric pressure at the airplane's flight altitude and comparing it to a preset pressure value. Air pressure decreases by about one-inch mercury for each 1,000-foot altitude increase.

Inside the instrument, the casing is a set of three aneroid wafers that are sealed but still able to expand and contract. These aneroid wafers are calibrated to sea level pressure of 29.92" mercury inside. An outside static pressure lower than 29.92" Hg (as experienced with a gain in altitude) causes the wafers to expand since the pressure inside of the sealed wafers is greater than on the outside. A higher static pressure causes the wafers to compress. When the static pressure increases or decreases, mechanical connections trigger the altimeter needle to show a corresponding altitude in feet.

The appearance of altimeters varies, but a common one is known as a three-point altimeter. This type of altimeter has a background similar to a clock with numbers from zero to 9 and three needles on the face. One is a short, wide needle that shows height in 10,000-foot increments, one is a slightly longer and wider needle that depicts height in 1,000-foot increments, and the longest needle shows height in 100-foot increments. Older altimeters have only one needle that circles once around the dial for every 1,000 feet in altitude.

Most altimeters in use today include a Kollsman window, which is an adjustable dial that allows the pilot to enter the local pressure values for his flight. Entering a pressure value in the Kollsman window adjusts the altitude for nonstandard pressure and gives a more accurate altitude.

Types of Altitudes

- Indicated Altitude:

  The altitude depicted on the altimeter when the pressure is set correctly in the Kollsman window.

- True Altitude:

  The height above sea level (MSL).

- Absolute Altitude:

  The height above ground level (AGL).

- Pressure Altitude:

  The altitude is shown on the altimeter when the standard atmosphere level of 29.92" Hg is entered in the Kollsman window, or the height above the standard datum plane. Pressure altitude is used often in flight planning calculations.

- Density Altitude:

  Pressure altitude adjusted for nonstandard temperature. Density often is described as how high the aircraft feels like it is since density altitude affects aircraft performance.

Altimeter Errors

Position

The position of static ports lends itself to disrupted airflow during certain maneuvers, phases of flight, and wind conditions. Disturbed airflow over the static port can cause erroneous readings on the altimeter.

Elasticity

Over time, the expansion and contraction of aneroid wafers in the altimeter can cause metal fatigue. Sometimes known as hysteresis, these changes in the elasticity of the instrument can cause inaccuracies.

Pilot

Pilots must establish the correct altimeter setting and enter it correctly into the Kollsman window in order for the altimeter to read correctly. Failure to set the altimeter correctly can cause altitude errors of hundreds of feet. A difference of 1" Hg can cause an altitude deviation of 1,000 feet.

Density

The density of the air changes from one area to the next, especially with temperature changes. Density errors associated with altimeters are apparent on longer flights, but also can happen on short flights that involve significant temperature changes.

A pilot will stay at the same height above the ground (as indicated on the altimeter) only if the temperature and pressure both remains the same. Flying from a high-pressure area to a low-pressure area without changing the altimeter would result in the aircraft being lower than expected. And because density changes with temperature, flying from a hot area to a cold area without changing the altimeter setting also will result in the aircraft flying a lower true altitude than expected.

Static Port Blockage

Blockage of the static port would result in static pressure being trapped inside of the instrument casing (but outside of the aneroid wafers), and the altimeter would freeze in place at the altitude it depicted at the time of the blockage. Since no air pressure changes would be measured, the altimeter needles would not move until the blockage was fixed.

Types of Altimeter

Barometric Altimeter

A barometric altimeter consists of a barometric capsule linked to a pointer by a suitable mechanical or electronic system. The pointer moves across the dial in response to changes in barometric pressure. The dial is calibrated in feet, or (less commonly) in metres.

Barometric altimeters are provided with a pressure setting control and sub-scale (Kollsman window) so that the altimeter may be calibrated according to the appropriate pressure setting to indicate flight level, altitude above mean sea level, or altitude above ground level.

The altimeter provides an output to the transponder system to enable the transmission of the flight level or altitude to the air traffic control.

Types of Barometric Altimeter

Barometric altimeter displays may be of two main types:

- Conventional analogue display;

- Electronic display.

The main types of conventional altimeter which have been used in aircraft are:

- Three-pointer altimeter;

- Drum-pointer altimeter;

- Counter-pointer altimeter;

- Counter drum-pointer altimeter.

The three-pointer altimeter and the counter drum-pointer altimeter are illustrated below:

It has been found that the displays of three-pointer altimeters, drum-pointer altimeters and counter-pointer altimeters are capable of being mis-read and several accidents have been attributed to this cause. Accordingly, counter drum-pointer altimeters are the the only type currently approved for use in commercial aircraft. In modern aircraft, conventional instruments are used mostly as standby instruments.

Modern aircraft are usually equipped with composite Electronic Flight Instrument System displays which combine the functions of several conventional instruments into one. Presentations vary according to the manufacturer's design philosophy. The illustration below shows a typical EFIS display in which the altitude is depicted on a vertical tape to the right of the attitude indicator. In the illustration, the altitude is 5100 ft. The altimeter pressure setting (29.92 in Hz) is depicted in green below the altitude tape.

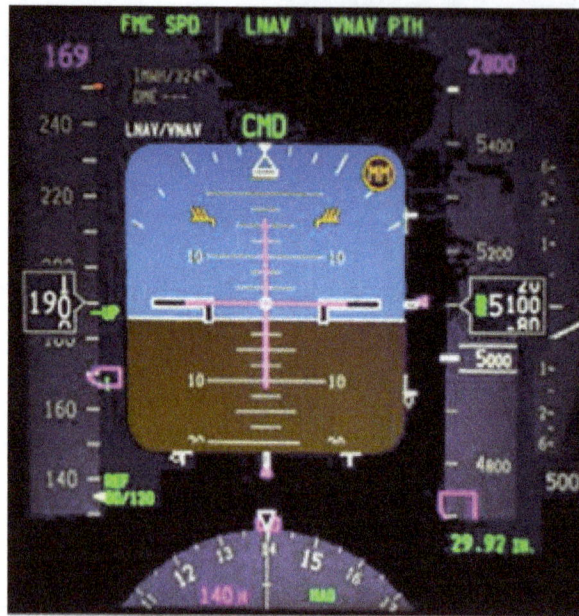

Radio Altimeter

A radio altimeter is an airborne electronic device capable of measuring the height of the aircraft above terrain immediately below the aircraft.

Early radio altimeters determined altitude by measuring the time between transmission of a radio signal from the aircraft and reception of the reflected signal. Modern systems use other means, for example, measurement of the change of phase between the transmitted and reflected signal.

In almost all cases, the display of radio height ceases when an aircraft climbs through 2500' above ground level (agl) and recommences when it descends through 2500' agl. This is confirmed visually by the appearance/disappearance of an 'OFF' flag and emergence of a pointer from behind a mask or activation of a digital display.

Radio altimeter calls may be either:

- Announced by one of the pilots, usually the Pilot Flying (PF) and Pilot Monitoring (PM), either in accordance with an SOPs or on an ad hoc basis.

- Generated automatically by a synthetic voice.

Many operators have an SOPs which requires a pilot call of "Rad Alt Live" to be made during descent as soon as practicable after height indications reappear at 2500 feet agl in order to enhance crew awareness of proximity to terrain.

Use of the radio altimeter is integral to both the function of ground proximity warning systems and to the operation of aircraft during Cat 2/3 approaches where it is used to determine the position of the aircraft in relation to the applicable decision height.

Prudent use of the radio altimeter can be a valuable defence against Controlled Flight Into Terrain (CFIT).

Airspeed Indicator

Airspeed indicator is instrument that measures the speed of an aircraft relative to the surrounding air, using the differential between the pressure of still air (static pressure) and that of moving air compressed by the craft's forward motion (ram pressure); as speed increases, the difference between these pressures increases as well.

Pressures are measured by a Pitot tube, a U-shaped apparatus with two openings, one perpendicular to the flow of air past the aircraft and one facing directly into the flow. Mercury or a similar liquid fills the bend in the tube, forming parallel columns balanced by the air pressure on each side. When static and ram pressure are equal, the columns have the same height. As the ram pressure increases, mercury on that side of the tube is pushed back and the columns become imbalanced. The difference between the two columns can be calibrated to indicate the speed; this value, called the indicated airspeed, may be given in knots, miles per hour, or other units.

Since the airspeed indicator is calibrated at standard temperature and pressure, its readings are inaccurate at different temperatures and altitudes. An (uncorrected) indicated airspeed is still used to estimate an aircraft's tendency to stall. Instruments that electronically correct for altitudinal differences and temperature give the true airspeed, which is used to calculate the aircraft's position. In faster aircraft, indicators that measure airspeed relative to the speed of sound, called Machmeters, are used.

The airspeed indicator works by comparing dynamic pressure—ram air pressure—and static pressure.

Each aircraft has specified airspeeds, of which the pilot needs to be aware. Pilots need to know the speeds at which an airplane will take off, land, stall, and otherwise operate safely during different phases of flight.

How it Works

The airspeed indicator is part of the pitot-static system, a differential pressure system that measures both dynamic air pressure from the pitot tube and static pressure from a static port. Inside the casing of the instrument is a sealed diaphragm that receives both static and dynamic pressure from the pitot tube.

Static pressure is also measured from inside the casing but outside of the diaphragm. The static pressures from both inside and outside of the diaphragm cancel each other out, leaving a measurement of total dynamic pressure, or ram air pressure.

As the airplane accelerates, the dynamic pressure from the pitot tube increases, causing the diaphragm to expand. Through mechanical linkage, this measurement of increased airspeed is shown in the airspeed indicator needle.

Markings and Limitations

Small, single-engine airplanes use color-coded airspeed markings to help the pilot operate safely and efficiently. These markings are helpful because they depict crucial aircraft speeds, known as V-speeds.

White Arc

The white arc on airspeed indicators depicts the normal flap operating range. Inside of the white arc, full flaps can be used. The top of the white arc indicates the highest speed at which flaps can be extended during flight, and operating at speeds outside of the white arc with flaps down can be unsafe.

Green Arc

The green arc on an airspeed indicator represents the normal operating range of the airplane.

Yellow Arc

The yellow arc is a cautionary range of airspeeds. It is advised that pilots only operate in the yellow arc in calm air. Flying at speeds in the caution range during turbulence can be unsafe.

Red Line

At the top of the yellow arc is a red line, which represents the maximum allowable airspeed for the airplane.

Airspeed Indicator Errors

An airspeed indicator will be ineffective if there is a blockage of the pitot tube or static port or both. A blockage is most commonly the result of insects, water, or ice.

If the pitot tube and its drain hole become blocked, the airspeed indicator acts like an altimeter, showing an increase in airspeed when the airplane climbs to a higher altitude and a decrease in airspeed during a descent.

If the pitot tube becomes blocked and the drain hole remains open, the ram air pressure will bleed out through the drain hole, leaving only static pressure in the pitot tube. The new static pressure in the pitot tube would be equivalent to the static pressure from the static port, and the airspeed indicator would read "0."

If the static port becomes blocked (but not the pitot tube), the airspeed indicator will work, but it will be inaccurate. Since the static air becomes trapped inside the casing at the altitude where the instrument stopped working, a climb will cause a lower than normal airspeed. When flying below the altitude at which the blockage occurred, the airspeed indicator will read higher than normal.

Emergency Operation

Some airplanes are equipped with pitot tube heating elements. Pitot heat is used as a preventative measure to stop ice from forming over the pitot tube and is activated when flying in cold weather.

Many small aircraft are equipped with an alternate static source that can be activated by pulling a lever in the cockpit in the event that a static port becomes blocked. The new alternate static pressure is a lower pressure than outside ambient pressure during flight, which results in slightly inaccurate instrument indications, but it provides a good enough indication to maintain positive aircraft control.

Vertical Speed Indicator

A Vertical Speed Indicator (VSI), also known as a Rate of Climb and Descent Indicator (RCDI) is an instrument which indicates the rate of climb or descent of an aircraft.

The VSI uses the aircraft pitot-static system to determine the vertical speed and depicts the result on a conventional needle and circular scale instrument, or on a ribbon at the side of an Electronic Flight Instrument System EADI.

Two typical VSI indications are depicted below. The first is a conventional indication while on the second instrument, the vertical speed is indicated on the scale at the extreme right of the instrument.

In a simple VSI, a barometric capsule is contained in a sealed case. The capsule is fed with static pressure from the pitot-static system, while the case is also connected to that system through a calibrated nozzle. The nozzle restricts the passage of air so that there is a time delay between a change in static pressure and that pressure being experienced within the case. Thus, if the aircraft climbs (or descends), the pressure within the capsule will decrease (increase) while that within the case will decrease (increase) at a lower rate due to the presence of the nozzle. Movement of the capsule is translated into movement of a needle by a mechanical system.

Conventional VSI.

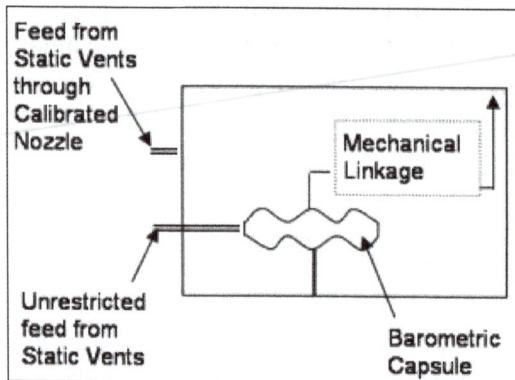

Schema of a simple Vertical Speed Indicator.

The vertical speed indicator is made up of a diaphragm inside of an airtight instrument casing. The diaphragm is connected by linkage and gears to the needle on the face of the instrument. Static pressure lines are connected to both the inside of the diaphragm and the instrument casing. The casing surrounding the diaphragm has a metered leak, which helps reflect the rate of climb or descent.

Pressure changes are measured instantaneously within the diaphragm as it expands and contracts from the pressure. The metered leak in the surrounding instrument casing also measures the pressure change, but the leak provides an intentional lag, allowing the instrument to measure the pressure change more gradually than inside the diaphragm. This lag comes from the consistent pressure leak and the corresponding rate of climb or descent as it's measured on the instrument needle in feet per minute. After a few seconds of level flight, the two pressures equalize and the vertical speed indicator shows 'o' feet per minute (fpm).

The result of a climb or descent is shown on the vertical speed indicator first as trend information (meaning a sudden climb or descent) and then shown as rate information, for example, 400 fpm.

Errors and Limitations

While the VSI is a crucial instrument for pilots, it has some limitations.

Turbulence

The vertical speed indicator is inaccurate during turbulence and when maneuvering abruptly. The lag involved with the calibrated leak is about six to eight seconds, rendering the vertical speed indicator almost useless when turbulence is encountered. If turbulence is encountered, the pilot should try to maintain an appropriate pitch attitude using the attitude indicator or outside visual references, instead of relying on a perfect instrument reading, sometimes called "chasing the needle," or trying to maintain a steady rate.

Static Port Blockage

If a static port becomes blocked, similar to an altimeter, the vertical speed indicator will indicate '0' and no change will be witnessed with either a climb or a descent. However, some airplanes are equipped with an alternate static source which provides an alternate source of static air to the flight instruments in the event of a blockage of the main static line.

Pitot-Static Probe

Pitot-Static probes measure total and static pressures and can be used to compute flow velocity.

The Pitot-Static probe is most commonly used in applications where only an air speed measurement is desired. Wind Tunnel facilities commonly use Pitot-Static probes to characterize their wind tunnels because the angle of the air flow is known.

The aviation and unmanned systems industries also use Pitot-Static probes to measure the aircraft's speed, Mach number, and altitude.

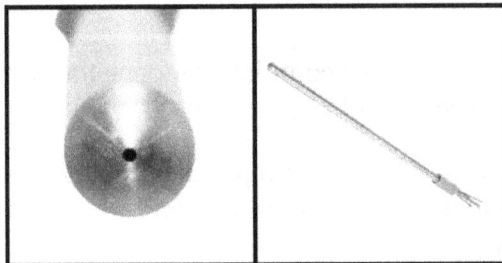

An aircraft pitot-static system comprises a number of sensors which detect the ambient air pressure affected (pitot pressure) and unaffected (static pressure) by the forward motion of the aircraft. These pressures are used on their own or in combination with each other to provide indications of various flight parameters.

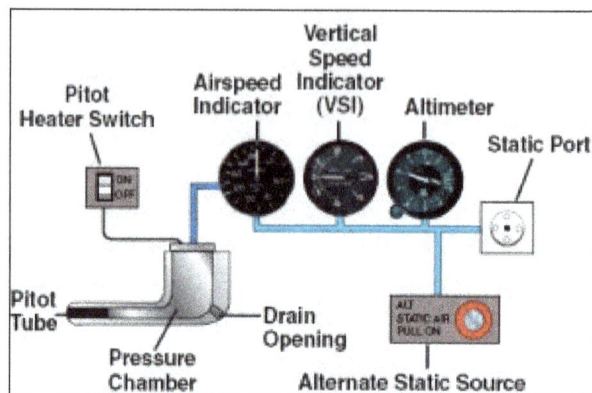

Diagram of a pitot-static system including the pitot tube, pitot-static instruments and static port.

These include:

- Altitude (Altimeter),

- Airspeed (Air Speed Indicator),

- Mach Number (Machmeter),

- Vertical speed (Vertical Speed Indicator).

Pitot and static pressure are also used in other equipment, such as the Autopilot and the Cabin Altimeter.

Static Pressure

Static pressure is measured through a number of vents, situated at aerodynamically neutral points on the aircraft fuselage. Vents are sited on either side of the fuselage and feed into a common tube; this has the effect of cancelling out to some extent errors arising from the position of the vents.

A combination of careful vent siting and accurate calibration reduces errors to an acceptable degree.

Commercial aircraft have at least two completely independent static systems to provide redundancy in the case of system failure.

Static vents are often plugged when the aircraft is parked for more than a short period of time to reduce the chance of blockage or contamination. Vents may be electrically heated to prevent blockage by ice.

Pitot Pressure

Pitot pressure is measured in a pitot tube or pressure head, which is an open tube facing forward along the axis of the aircraft. The pressure measured in the tube is a combination of static pressure and pressure due to the aircraft forward speed. Pitot tubes are carefully sited to reduce to a minimum error due to the flow of air over the aircraft.

Commercial aircraft have at least two completely independent pitot systems to provide redundancy in the case of system failure.

Pitot tubes are normally covered when the aircraft is parked for more than a short period of time to reduce the chance of blockage or contamination. They are invariably electrically heated to reduce contamination by moisture and prevent blockage by ice.

Air Data Computer

Most modern aircraft are fitted with an Air Data Computer (ADC). This computer uses inputs from the pitot-static system and from temperature sensors to determine Indicated Airspeed, Mach Number, True Airspeed, Altitude, Vertical Speed, Outside Air Temperature (OAT) and Total Air Temperature (TAT). These data are fed to aircraft systems, especially the Electronic Flight Instrument System.

Turn and Bank Indicator

Turn and bank indicator is an aircraft instrument containing one indicator to show turning, or rotation about the vertical axis, and another to show banking, or rotation about the longitudinal axis. The two indicators are essentially separate instruments, but they are customarily placed together. The bank indicator is the simpler of the two and consists of a curved glass tube filled with a damping liquid in which a small steel ball rolls. When the craft is horizontal, the ball is located in the lowest part of the tube; as the craft banks, gravity holds the ball at the lowest point as the tube rotates from side to side. The tube can be calibrated to show the angle of banking. The turn indicator contains a gyroscope that develops a torque when the craft rotates. This torque controls a pointer that indicates to the pilot in degrees per unit of time the rate at which the craft is turning.

Principle Of Turn Indicator

Uses single axis rate gyroscope. As the aircraft turns, the rotor is subjected to a force F. the direction of precession (P) is moved 90 degrees ahead in direction of the rotor spin, causing the rotor to tilt about the Y1 and causing the spring to be extended. Equilibrium occurs when the spring has equal force when compared with the turning force. Precession is proportional to the rate of turn. No erection will be required however; the rotor speed has to be kept constant at 4,000 – 5,000 rpm. When the speed is too slow, the precession will be lesser than normal resulting in an under reading.

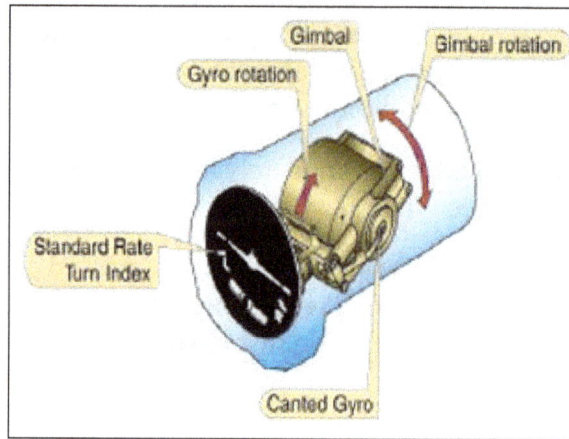

The Turn Coordinator detects a turn in 2 axes, which is based on the gimbal rotation and the gyro rotation. The Gimbal frame is angled at 30 degrees from the vertical axis. During a turn the initial rate of banking will be shown and once it stabilizes, the rate of the turn will be shown.

References

- Altimeter, encyclopedia: nationalgeographic.org, Retrieved 03 January, 2019

- Basic-flight-instruments-the-altimeter-282608: hebalancecareers.com, Retrieved 25 February, 2019

- Radio-Altimeter: skybrary.aero, Retrieved 23 April, 2019

- Airspeed-indicator, technology: britannica.com, Retrieved 16 June, 2019

- Basic-flight-instruments-the-airspeed-indicator-282607: ebalancecareers. com, Retrieved 25 January, 2019

- Vertical-Speed-Indicator: skybrary.aero, Retrieved 14 July, 2019

- Turn-and-bank-indicator, encyclopedia-science-tech-aviation-instruments: infoplease.com, Retrieved 16 May, 2019

Permissions

Index